AAPS Advances in the Pharmaceutical Sciences Series

Volume 60

AAPS and Springer Nature Seek New Editor Advances in Pharmaceutical Sciences Book Series

Springer Nature, the American Association of Pharmaceutical Scientists, are pleased to announce an exciting opportunity for an exceptional candidate to serve Editor for the Advances in Pharmaceutical Sciences Book series.

Interested candidates should provide a statement of why they are interested in this position and what their vision would be for the editorship of this series, along with two or three references and/or letters of recommendation from knowledgeable colleagues and a current CV to Publications Manager, Matt Baughman (BaughmanM@aaps,org).

Review the job description (PDF) before applying.

Applications Due: May 1, 2024

Volunteer Opportunities - American Association of Pharmaceutical Scientists (aaps.org)

Bhavishya Mittal

Sustainable Global Health Systems and Pharmaceutical Development

Springer

aaps

Bhavishya Mittal
Vice President, Product Development
and Manufacturing
BioDuro-Sundia
Irvine, CA, USA

ISSN 2210-7371 ISSN 2210-738X (electronic)
AAPS Advances in the Pharmaceutical Sciences Series
ISBN 978-3-031-50414-3 ISBN 978-3-031-50415-0 (eBook)
https://doi.org/10.1007/978-3-031-50415-0

This Springer imprint is published by the registered company Springer Nature Switzerland AG
The registered company address is: Gewerbestrasse 11, 6330 Cham, Switzerland

बड़ा हुआ तो क्या हुआ जैसे पेड़ खजूर |
पंथी को छाया नहीं फल लागे अत दूर ||
– कबीर

Humility is the solid foundation of all virtues
– Confucius

Preface

If there is any one secret to success, it lies in the ability to get the other person's point of view and see things from his angle and your own. (Henry Ford)

The importance of the pharmaceutical industry for human society was never in question. However, the clear and present danger of the COVID-19 pandemic brought our industry into sharp focus. The world recovered from the pandemic due to the significant efforts of numerous research institutes, pharmaceutical companies, regulatory agencies, government institutions, and many other healthcare agencies. Such a recovery effort has never happened in human history. However, the post-pandemic years have also presented unique opportunities to discuss how society can capitalize on this innovative industry to deliver the safest and most cost-effective medicines for the myriad of diseases.

For me personally, it all started with a seemingly benign question from a close family member, "Why do pharmaceutical products cost so much?". I knew the answer and tried to frame it in the context of the high costs of innovation and risks associated with product development. However, the second question was more profound for me: "What is being done to reduce the costs of pharmaceutical products?". The answer to this question was not as simple and did require me to reflect on the question a lot more. It made me wonder how the public perceives the innovation potential of the pharmaceutical industry and why all conversations around pharmaceuticals revolve around the costs. I also realized that being a member of the pharmaceutical industry for 21+ years may have insulated me from the perception of the industry in the general public.

There is no doubt that the pharmaceutical industry has done so much good for the world (and continues to do so). But as a responsible member of the industry, I also believe that it is my solemn duty to understand and appreciate the perspectives of the numerous stakeholders and patients for whom we are developing these products in the first place. Such introspective exercises help build empathy and create safe-space conversations aiming to appreciate other people's viewpoints. The quest to answer these questions led to reviewing numerous topics on global health inequality, healthcare technology assessment, stakeholder analysis, pharmaceutical sciences, and the latest trends, such as Industrial Revolution 4.0. The results of this research are amalgamated in this book.

More than 750 companies are listed as either biotechnology or major pharmaceutical companies on the NYSE and NASDAQ combined. These companies cover a whole spectrum of business-operating models all the way from fully integrated research companies with thousands of employees to completely virtual entities that operate with minimal permanent staff. No matter what their *modus operandi* may be, all these companies are developing new drug products for a vast number of therapeutic indications. From a societal viewpoint, with the amazing number of resources spent on each new drug candidate, the total costs start to add up quickly! This further necessitates that any efforts that can help to lower the development time and costs should be carefully reviewed and, if possible, integrated into the product development strategy.

Pharmaceutical innovation without integrating economics into product design is an inherently futile and non-sustainable proposition. Therefore, the time to discuss how the development of robust and productive healthcare options can be achieved economically has now come, for, after all, no industry is immune to the laws of economics. With this premise, this book hopes to spur a conversation that aims to integrate drug development, economics, and efficient manufacturing in a fashion that improves the accessibility of drugs to patients. Such conversations allow pharmaceutical scientists to get a step closer to building an "empathetic bridge" and be more aware of the patient's needs.

As the eminent economist Milton Friedman once said, "Only a crisis—actual or perceived—produces real change. When that crisis occurs, the actions taken depend on the ideas lying around." In that spirit, ideas to improve healthcare systems are regularly discussed in the literature. However, this book hopes to integrate the opportunities within pharmaceutical R&D with the sustainability of healthcare systems. Therefore, all this book is doing is collating those ideas and thus offering a perspective that summates economics, manufacturing sciences, and pharmaceutical R&D together. It is my sincere hope that the readers find that this book is thoroughly researched, its approach well-thought-out, and is a manuscript worthy of their time and consideration.

Irvine, CA, USA
Bhavishya Mittal

Acknowledgments

I would like to thank my mentors, Dr. Michael Levin and Dr. Ajaz Hussain, for listening to my ideas and guiding me toward various critical source materials. I also sincerely thank the Acquisitions, Editorial Project Management, and Publishing teams at Springer. In particular, I appreciate the help from Ms. Merry Stuber and Ms. Charlotte Nunes that made this project come to fruition.

I am also thankful to my parents (Dr. J.P. Mittal and Madhu Mittal), who have positively influenced my life and have always provided their unbridled support, blessings, and encouragement. I am incredibly thankful to my loving wife, Shalini, for her unconditional love, positive attitude, and constant reassurance, which helped me to complete this project on time. I thank my son, Kern, for patiently proofreading this manuscript and providing critical input. I am also thankful to my daughter, Ariana, for their perennial encouragement and understanding while I worked on this book.

Contents

List of Figures

List of Tables

About the Author

Bhavishya Mittal is the Vice President of Product Development and Manufacturing at BioDuro-Sundia (Irvine, CA), where he leads the strategy and operations for drug product development of client-owned assets within the portfolio. Before BioDuro-Sundia, Bhavi worked at Kronos Bio (San Mateo, CA), AbbVie (Irvine, CA), US FDA (Silver Spring, MD), Takeda Pharmaceuticals (Cambridge, MA), and Astellas Pharmaceuticals (Norman, OK). Bhavi is also an adjunct faculty at the Keck Graduate Institute's Henry E. Riggs School of Applied Life Sciences (Claremont, CA). He has a PhD in Materials Engineering from the Pennsylvania State University and a BS in Chemical and Bio Engineering from the National Institute of Technology, Jalandhar (India).

Bhavi has 21+ years of industrial experience in the formulation design, process development, and project management of numerous drug products of small therapeutic molecules (oncology, inflammation, GI, and CNS indications) aimed for NDA and ANDA filings. His expertise includes product development strategy, project management, outsourcing management, phase-appropriate formulation design (immediate and modified release), process engineering, statistical process control, QbD implementation, FMEA/risk mapping, data analytics, and process modeling/simulation of unit operations for solid oral dosages.

An avid sports enthusiast and a history buff, Bhavi is a passionate supporter of continuing education and firmly believes in the noble tradition of passing-on knowledge to younger generations. He loves to travel, learn about various cultures, and spend time with his family.

Abbreviations

AAC	Actual Acquisition Cost
AAM	Association for Accessible Medicines
ADC	Antibody Drug Conjugate
ADME	Absorption, Distribution, Metabolism and Excretion
AHRQ	Agency for Healthcare Research and Quality
AI	Artificial Intelligence
AIDS	Acquired Immune Deficiency Syndrome
AM	Additive Manufacturing
AMP	Average Manufacturer Price
AMR	Antimicrobial Resistance
ANDA	Abbreviated New Drug Application
ARV	Antiretrovirals
ASQ	American Society for Quality
ATMP	Advanced Therapy Medicinal Products
AWP	Average Wholesale Price
BCS	Biopharmaceutical Classification System
BEA	Bureau of Economic Analysis
BLA	Biologics License Application
BPCIA	Biologics Price Competition and Innovation Act
CDER	Center for Drug Evaluation and Research
CET	Cost-Effectiveness Threshold
CHE	Current Health Expenditure
CHIP	Children's Health Insurance Program
CHO	Chinese Hamster Ovary
CL	Compulsory Licenses
CM	Continuous Manufacturing
CMC	Chemistry, Manufacturing, and Control
CMS	Centers for Medicare and Medicaid Services
COG	Cost of Good
COPQ	Cost of Poor Quality
CPP	Critical Process Parameter

CQA	Critical Quality Attribute
CSDH	Commission on Social Determinants of Health
DAH	Development Assistance for Health
DALY	Disability-Adjusted Life Year
DCS	Developability Classification System
DDD	Daily Defined Dose
DMPK	Drug Metabolism and Pharmacokinetics
EM	Essential Medicines
FDA	US Food and Drug Administration
FDC	Fixed-Dose Combination
FDCA	Food, Drug, and Cosmetic Act
GATT	General Agreement on Tariffs and Trade
GDP	Gross Domestic Product
GHG	Global Health Governance
GHO	Global Health Observatory
HAQ	Healthcare Access and Quality
ICER	Incremental Cost-Effectiveness Ratio
IGO	Inter-Governmental Organization
IHI	Institute for Healthcare Improvement
IHME	Institute for Health Metrics and Evaluation
IND	Investigational New Drug
IP	Intellectual Property
IPCC	Intergovernmental Panel on Climate Change
LCM	Life-Cycle Management
M&S	Modeling and Simulation
mAb	Monoclonal Antibody
MCB	Master Cell Bank
MDG	Millennium Development Goal
NCE	New Chemical Entity
NDA	New Drug Application
NDC	National Drug Code
NGO	Non-Governmental Organization
NTD	Neglected Tropical Diseases
OBA	Outcomes-Based Agreement
OBD	Optimal Biological Dose
OECD	Organization for Economic Co-operation and Development
OTC	Over-the-Counter
P&R	Pricing and Reimbursement
PAT	Process Analytical Technology
PBM	Pharmacy Benefit Manager
PCE	Personal Consumption Expenditures
PDM	Product Design Management
PDP	Prescription Drug Plan
PDUFA	Prescription Drug User Fee Act
PFDD	Patient-Focused Drug Development

PhRMA	Pharmaceutical Research and Manufacturers of America
PHSA	Public Health Services Act
PPF	Production Possibilities Frontier
PPP	Public-Private Partnerships
PROM	Patient-Reported Outcome Measure
PSO	Patient Safety Organization
QALY	Quality-Adjusted Life Year
QbD	Quality by Design
QbED	Quality by Efficient Design
QTPP	Quality Target Product Profile
R&D	Research and Development
SDG	Sustainable Development Goals
SOC	Standard of Care
SPP	Special Patient Populations
THE	Total Health Expenditure
TPP	Target Product Profile
TRIPS	Trade-Related Aspects of Intellectual Property Rights
TTM	Time-to-Market
UN	United Nations
UNICEF	United Nations Children's Fund
USAID	US Agency for International Development
USPTO	United States Patent and Trademark Office
VA	Department of Veterans Affairs
VBP	Value-Based Pricing
WCB	Working Cell Bank
WHO	World Health Organization
WAC	Wholesale Acquisition Cost
WTO	World Trade Organization

Chapter 1
The Need to Introspect

No institution should expect to be free from the scrutiny of those who give it their loyalty and support.

Queen Elizabeth II

Contents

Abstract Global health inequity is a significant crisis facing humankind and is a gaping hole in our efforts to achieve a just and equal global society. Significant efforts have been taken by numerous political leaders, visionaries, and international organizations over many generations to outline the various actions, approaches, and recommendations that aspire to close the health inequity gap. Unfortunately, despite of well-intentioned efforts from stakeholders, access to medicines remains a vexing issue. Potential solutions include responsible governance, positive cultural shifts, balanced socioeconomics, infrastructure growth, and increased public awareness. Another approach could be to improve the efficiency of the pharmaceutical development process. From a for-profit company's perspective, substantial financial risk must be undertaken without any assurance of economic reward as pharmaceutical research is complicated, time-consuming, attritive, and costly, with estimated

B. Mittal, *Sustainable Global Health Systems and Pharmaceutical Development*, AAPS Advances in the Pharmaceutical Sciences Series 60, https://doi.org/10.1007/978-3-031-50415-0_1

pre-approval costs of $2.6 billion per new drug. Drugs that are successful in the marketplace are typically the only sources of revenue for a pharmaceutical company, whether big or small. However, there is an opportunity to reduce the development cost for new drug products. To stem the decline in productivity and to optimize its resources, a desirable future state for the pharmaceutical industry integrates economics into product design and development. Such an approach will allow for the decision-making to be parameterized, cost of goods to be lowered, wastage reduced, patient-centricity built into the design, and manufacturing/distribution efficiencies to be gained. The financial benefits of such an approach could allow these savings to be passed on to the stakeholders, improve the value proposition of pharmaceuticals, help bridge the health inequity gap, and enhance the sustainability of our global healthcare systems. Such a utopian future requires understanding where do we stand with our healthcare systems today.

Keywords Global healthcare · Health inequity · Access to medications · Sustainability · TRIPS · Healthcare expenditures · Pharmaceutical spending · Product development

1.1 A Call for Introspection

The Oxford English Dictionary defines introspection as the "action of looking within, or into one's mind; examination or observation of one's thoughts and feelings". Introspection relies on observing one's mental state; in a spiritual context, it may refer to examining one's soul. Since the inception of psychology as an academic discipline, scholars have extensively used first-person (or introspective) inquiry: it is applied during theory development and hypothesis formulation; it permeates the data-collection stage and impacts interpretation and theory development. In all these cases, scientists' expertise, understanding, perspective, or expectations inform their scholarly practice (Weger et al., 2019). Introspection requires time, an honest effort, and a willingness to face facts, even if they might be unpleasant or unexpected. Without the emotional intelligence and mental acuity to conduct an introspection and be willing to scrutinize one's evolution, no improvement can be made that is well-intentioned, progressive, sustainable, and inspirational. Like people, organizations also need to conduct such contemplative investigations with the hope of providing balance to competing priorities. Often, in moving an organization's business forward, the original calling of doing something novel can sometimes get lost in the din and banality of mundane daily activities. Introspection gives an opportunity to pause and to reinvigorate by connecting back to that emotional and motivational touchstone. It also hardens the resolve to bring improvement and change. Cold-hard facts and data help in such exercises.

One particular tool that can help in these introspective exercises is benchmarking. The American Society for Quality (ASQ) defines benchmarking as the quantitative exercise of measuring products, services, and processes against those of

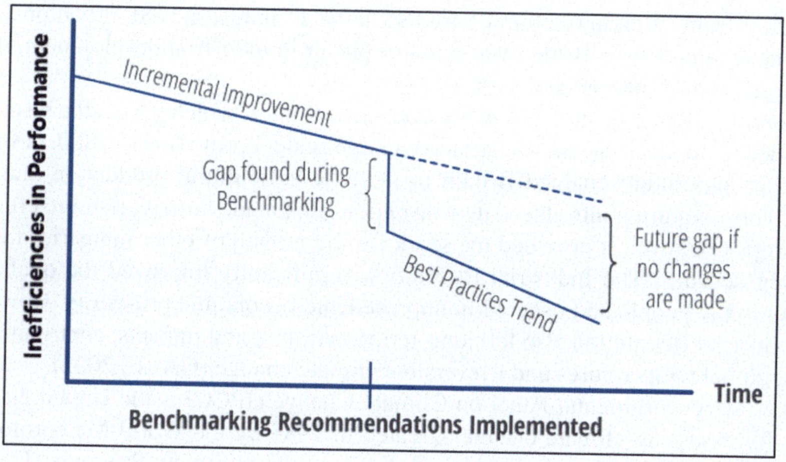

Fig. 1.1 Incremental quality improvement vs. benchmarking breakthroughs (adapted from ASQ, 2024)

organizations known to be leaders in one or more aspects of their operations. Benchmarking provides necessary insights to help researchers understand how an organization compares with similar organizations, even if they are in a different business or have a different group of customers. Benchmarking can also help organizations identify areas, systems, or processes for improvements—either incremental (continuous) improvements or dramatic (business process re-engineering) improvements (ASQ, 2024). This concept is highlighted in Fig. 1.1, where hypothetical inefficiencies in performance are plotted as a function of time. Every organization and workflow undergoes incremental improvements with time. However, when such exercises are performed, and the underlying recommendations are implemented, the performance inefficiencies can be significantly reduced, as highlighted by the "Best Practices Trend" line in Fig. 1.1. Therefore, benchmarking strategies help identify potential areas of improvement and help any organization strive toward increased efficiencies and higher productivity. One such introspective exercise that is key to our survival is climate change and how we as a species can address it.

1.1.1 *Philosophical Digression: Urgency of Addressing Climate Change*

History is full of historiographical landmarks. These landmarks are elaborate and ideal constructions that give structure and coherence to our historical narratives and define the significant research questions. One of the most important landmarks in history is the 18th century Industrial Revolution, which marked the onset of industrial society and defined the critical mechanisms of its progress (De Vries, 1994). The Industrial Revolution was a significant turning point in history, marking the transition from rural, agricultural-based societies to industrialized, urban ones. By the nineteenth century, Britain was a major player in industrial development, holding large reserves of coal and iron.

Moreover, Britain controlled many overseas colonies, which gave it the resources it needed to conduct industrialization on a large scale. From 1600 to 1800, developments in agriculture enabled Britain to increase agricultural production dramatically. This revolution introduced new inventions that made farm work more efficient and crops bountiful. It provided the spark for the growth of other industries in succeeding decades. The Industrial Revolution significantly improved the quality of life for many people and ushered in unprecedented economic prosperity. However, the Industrial Revolution also left long-term environmental impacts, contributing to rising global temperatures and irreversible climate change (GALE, 2022).

The Intergovernmental Panel on Climate Change (IPCC) is the United Nations body for assessing climate change. Created in 1988 by the World Meteorological Organization (WMO) and the United Nations Environment Program (UNEP), IPCC's objective is to provide governments at all levels with scientific information that they can use to develop climate policies. IPCC reports are also vital inputs into international climate change negotiations. In its latest assessment report, IPCC solemnly observes that human activities, principally through emissions of greenhouse gases, have unequivocally caused planetary warming, with global surface temperature reaching 1.1 °C above 1850–1900 in 2011–2020. Global greenhouse gas emissions have continued to increase, with unequal historical and ongoing contributions arising from unsustainable energy use, land use and land-use change, lifestyles, and consumption and production patterns across regions, between and within countries, and among individuals. Widespread and rapid changes in the atmosphere, ocean, cryosphere, and biosphere have occurred. Human-caused climate change already affects many weather and climate extremes in every region. This has led to widespread adverse impacts and related losses and damage to nature and people. Vulnerable communities that historically contributed the least to climate change are disproportionately affected.

There is no doubt that climate change is a threat to human well-being and planetary health. However, the IPCC estimates a rapidly closing window of opportunity to secure a livable and sustainable future for all. Climate resilient development is implementing greenhouse gas mitigation and adaptation measures to support sustainable development for all. It is enabled by increased international cooperation, including improved access to adequate financial resources, particularly for vulnerable regions, sectors, and groups, and inclusive governance and coordinated policies. There are multiple opportunities for scaling up climate action, as shown in Fig. 1.2. For example, numerous options to reduce industrial emissions will entail coordinated action throughout value chains to promote all mitigation options, including demand management, energy and materials efficiency, circular material flows, abatement technologies, and transformational changes in production processes.

The IPCC report points out that limiting human-caused global warming requires net zero carbon dioxide emissions. There is no exclusive solution, so any efforts to combat climate change require a multifaceted approach. Adaptation and mitigation

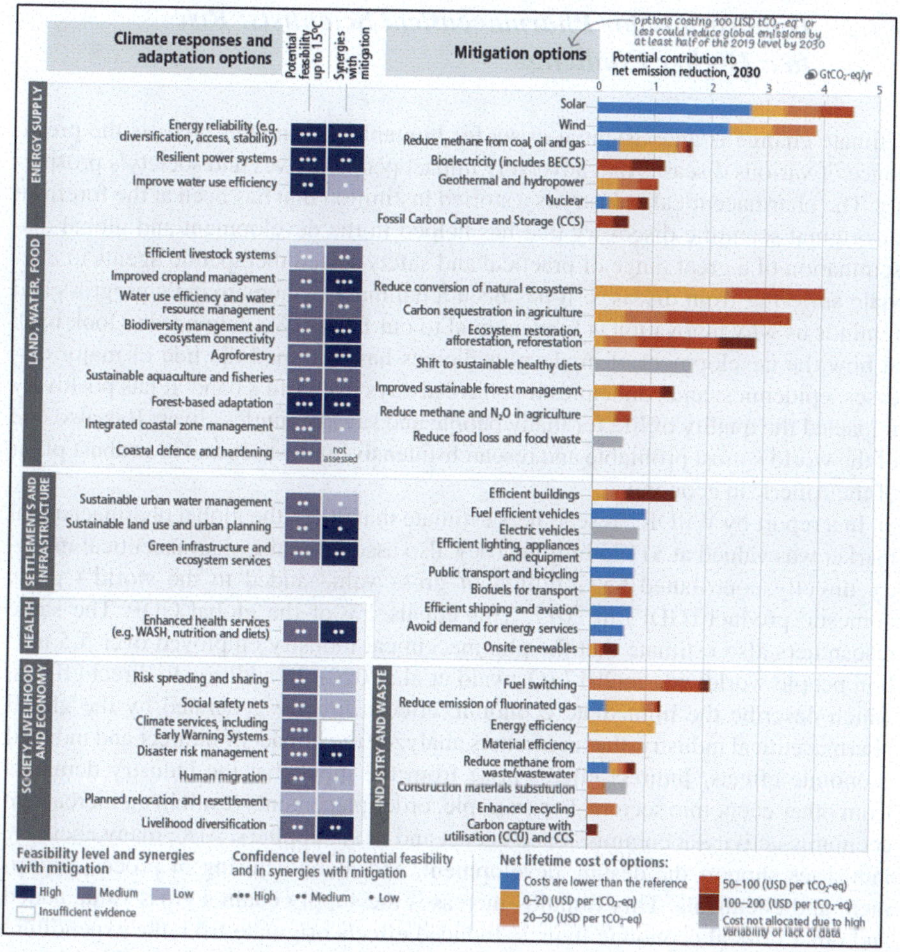

Fig. 1.2 Feasibility of climate responses, adaptation, and potential mitigation options (IPCC, 2023)

actions across scales, sectors, and regions prioritizing equity, climate justice, rights-based approaches, social justice, and inclusivity lead to more sustainable outcomes, reduce trade-offs, support transformative change, and advance climate-resilient development. Deep, rapid, and sustained mitigation and accelerated implementation of adaptation actions in this decade would reduce projected losses and damages for humans and ecosystems and deliver many co-benefits, especially for air quality and health. Delayed mitigation and adaptation action would lock-in high-emissions infrastructure, raise risks of stranded assets and cost-escalation, reduce feasibility, and increase losses and damages. Near-term actions involve high up-front investments and potentially disruptive changes that can be lessened by a range of enabling policies. The choices and actions implemented in this decade will have impacts now and for thousands of years (IPCC, 2023).

1.1.2 *Opportunity for Pharmaceutical Scientists: Rise to a Higher Challenge*

Climate change is an existential crisis for humankind. Another crisis is the prevalence of various diseases that adversely impact people's lives and society's prosperity. The pharmaceutical industry is a storied institution that has been at the forefront of original scientific discovery that has helped in the development and global dissemination of a great range of practical and safety-tested therapeutic agents to alleviate suffering from diseases. It has been a hallmark of humankind's progress and reminds us why innovation is fundamental to our future. We only need to look back at how the development of modern medicines has stemmed the tide of major diseases, epidemics, and pandemics that have always been life's bane. It has positively impacted the quality of life for many people and saved countless lives. It is also one of the world's most profitable and research-intensive industries and is a robust pillar of the American economy.

In a report by WifOR,[1] researchers estimate that 2017, the global pharmaceutical market was valued at $1135 billion. They also assert that the pharmaceutical industry directly contributed $532 billion of gross value added to the world's gross domestic product (GDP) in 2017. This equals 1% of the global GDP. The same researchers also estimate that the pharmaceutical industry employed over 5.5 million people worldwide in 2017 (Ostwald et al., 2020). In addition to direct effects, which describe the immediate economic effects directly generated by the global pharmaceutical industry, the researchers analyzed the impact of indirect and induced economic effects. Indirect effects arise from the input that the industry demands from other economic sectors. For example, order placements result in an increase in economic activity at commissioned agents and their suppliers. Also, many ancillary industries support the design, development, and manufacturing of process equipment and chemicals. This stimulus increases the supply chain's gross value added and other critical economic figures. Induced effects originate from the expenditure of directly and indirectly generated incomes and the concomitant increase in demand. The combination of indirect and induced effects is called spillover effects. Total economic effects referring to the sum of all three (direct, indirect, and induced effects) can be visualized as shown in Fig. 1.3. Through its economic activity, the global pharmaceutical industry contributes to the UN Sustainable Development Goal 8: Promote economic growth, employment, and work for all (Ostwald et al., 2020) (*please see* Fig. 2.7 *for a detailed discussion on Sustainable Development Goals*).

Similar economic impact analyses have been done by other organizations as well. Pharmaceutical Research and Manufacturers of America (PhRMA) is a

[1] WifOR is an independent economic research institute that originated from a spin-out of the Department of Public Economics and Economic Policy at the Technical University of Darmstadt, Germany. WifOR's fields of research include Economic, Environmental and Social Impact Analyses as well as Labor Market and Health Economy research.

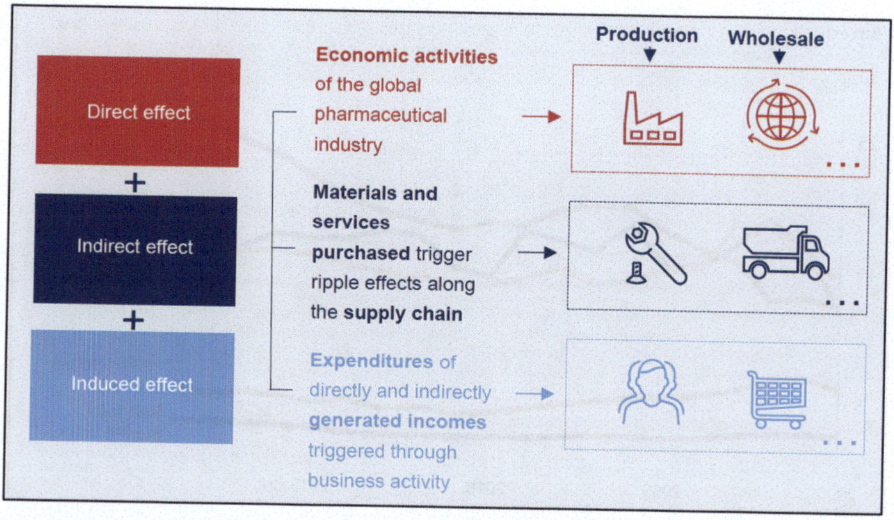

Fig. 1.3 Economic impact analysis through activities of the global pharmaceutical industry (Ostwald et al., 2020)

business organization that represents the United States' leading innovative biopharmaceutical research companies. Formed in 1958, PhRMA promotes public policy that supports medical research to address patients' needs. As per a policy paper published by PhRMA in 2017, there are about 7000 medicines in development, with about three-quarters having the potential to be first-in-class treatments. According to the same policy paper, investments in biopharmaceutical innovation improve and save lives, drive tremendous contributions to the American economy, and solidify America's role as a leader in medical innovation. Biopharmaceutical companies invest about $75 billion annually in Research & Development (R&D) in the United States, more than any other industry, and support nearly 4.7 million jobs (PhRMA, 2017). Pharmaceutical companies have devoted a growing share of their net revenues to R&D activities, averaging about 19% over the past two decades. By comparison, other research-intensive industries, like software and semiconductors, averaged about 15%, as shown in Fig. 1.4 (CBO, 2021).

Total annual healthcare expenditures in the United States, and the change from year to year, can be influenced by many different factors. Economic disruptions, population demographics, technology and healthcare practice changes, and evolving disease patterns all play a role. Rarely does a single disease have a significant impact. An exception is the COVID-19 pandemic. Among its many effects, the pandemic increased US healthcare expenditures by an unprecedented magnitude not experienced since 2002. In 2020, national spending on healthcare services grew by 9.5% to $4.1 trillion, while GDP dropped by nearly 20%. Many different factors also influence annual changes in prescription drug expenditures.

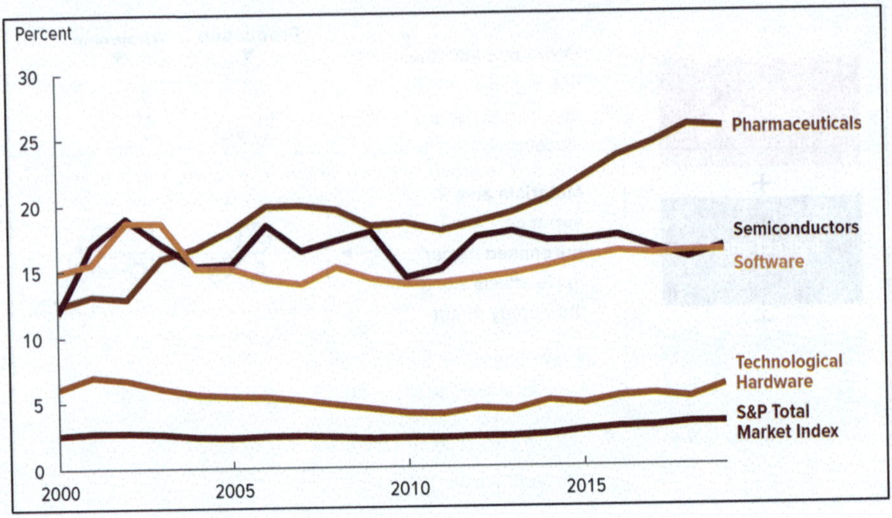

Fig. 1.4 Average R&D intensities for publicly traded US companies (CBO, 2021)

Overall, pharmaceutical expenses in the US grew 7.7% in 2021, for a total of $576.9 billion (Tichy et al., 2022). The top 25 drugs by expenditures for the overall US market in 2021 are shown in Table 1.1. Regarding drug spending by therapeutic class in the US, new or improved specialty drugs for diabetes, various cancers, autoimmune disorders (such as rheumatoid arthritis or multiple sclerosis), and HIV have propelled large retail spending increases for those illnesses (Fig. 1.5). New drugs can significantly increase retail spending because they have higher prices, are in high demand, or both. Spending decreases can result when patent protection expires on leading drugs, and low-cost generic versions are introduced (CBO, 2021). From an economics perspective, all of these are vital statistics and are indicative of the impact of the pharmaceutical industry on society. However, despite its excellent track record, the pharmaceutical industry has a significant image problem (*please see* Sect. 8.3 *for a detailed discussion on the public perception of the industry*). Therefore, the industry could benefit from a reflective assessment that evaluates its business model, its challenges, and the opportunities it can take advantage of to improve its value in the eyes of its numerous stakeholders. However, before such a contemplative exercise can be undertaken, it is critical to assess the current healthcare landscape of our global society.

Table 1.1 Top 25 US drugs by overall expenditures in 2021 Tichy et al. (2022) (Reprinted with permission from Oxford University Press)

Drug[a]	2021 expenditures ($ thousands)	Percent change from 2020
Adalimumab	28,498,153	14.6
Apixaban	15,795,250	23.2
Dulaglutide	12,198,284	40.2
Semaglutide	10,718,460	90.1
Ustekinumab	10,681,921	28.3
Insulin glargine	10,027,314	3.2
Pembrolizumab	9,860,807	18.7
Bictegravir/emtricitabine/ tenofovir alafenamide	9,570,249	19.0
Etanercept	8,094,614	4.2
Empagliflozin	8,022,438	40.8
Rivaroxaban	7,135,012	7.5
Sitagliptin	6,406,544	2.4
Insulin aspart	6,043,212	2.1
Insulin lispro	5,700,806	2.7
Dupilumab	5,604,661	55.2
Immune globulin	5,371,428	6.7
Secukinumab	4,613,406	3.9
Liraglutide	4,546,309	−3.7
Epinephrine	4,382,123	13.9
Nivolumab	4,222,753	6.8
Ibrutinib	4,163,655	8.1
Budesonide/formoterol	4,155,298	0.1
Ocrelizumab	4,132,431	16.7
Lisdexamfetamine	4,100,293	8.9
Lurasidone	4,025,640	7.4

[a]For each drug listed, the expenditures shown are the total for branded and generic products (including biosimilars) and of various dosage forms unless otherwise stated

1.2 Health Inequity: A Somber Reminder

Good health is one of the biggest blessings. However, it is also sad that disease has been an integral part of life. From time immemorial, disease has caused incalculable pain, dysfunction, distress, social problems, and death. It has also altered the course of human history. Some of the worst epidemics and pandemics in history have doomed whole civilizations and brought once-powerful nations to their knees,

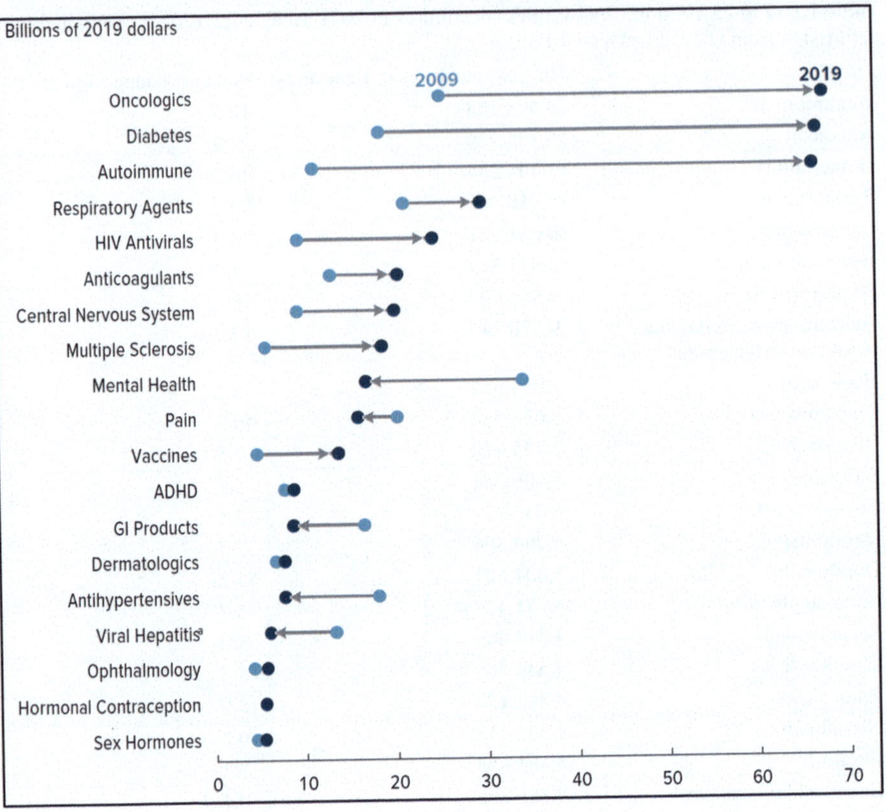

Fig. 1.5 Total US retail drug spending (in billions of 2019 dollars) by therapeutic class in 2009 and 2019 (CBO, 2021)

killing millions. Significant examples include the bubonic Plague of Justinian (541–542 AD) that ravaged the Byzantine Empire and led to its decline, the Black Death (1346–1353 AD) that wiped out over half of Europe's population, or the post-Columbian New World Plagues of the late 15th and 16th centuries that killed nearly 90% of the indigenous population in the Western Hemisphere (Jarus, 2022). These terrible disease outbreaks still threaten humanity, like the COVID-19 pandemic, which killed more than 5 million people worldwide from January 2020 to December 2022 (WHO, 2023b). However, thanks to the advances in epidemiology, we no longer face the same extremely dire consequences as our ancestors once did. Nevertheless, one should never underestimate the power of diseases. Such hubris will only lower our guard down and is the recipe for doom!

The development of the vaccine for the COVID-19 pandemic is an interesting case study. The steps that produced the most rapid vaccine rollout in history were only possible due to extraordinary efforts of basic scientific research and synergistic collaboration between the pharmaceutical industry, numerous research scientists, healthcare providers, and the federal government. To offset the cost of developing the COVID-19 vaccine and to help mobilize approved vaccines as quickly as possible to the American public, the government established nearly $10 billion in federal funds. This has dramatically accelerated the timeline for the development of vaccines through clinical trials, regulatory review, and mass distribution of a vaccine (Caldwell, 2021).

Vaccine development is typically measured in years, not months. However, as the COVID-19 pandemic raged in 2019–2020, scientists worked hard to use mRNA and vector-based platforms to rapidly develop safe, effective, low-cost vaccines in record time. In terms of numbers, as of 22 October 2023, more than 13 billion vaccine doses have been administered worldwide (WHO, 2023b). Such speed of vaccine development and deployment is genuinely unprecedented. The rapid response of the government and the pharmaceutical industry has also allowed us to evaluate what else can be done to accelerate development efforts to produce safe, effective, and cost-effective drugs. After all, a pandemic is not the only health emergency, and we should not have to wait for the next one to mobilize the might of our entire scientific, regulatory, and business acumen!

1.2.1 Statistics of Global Health

In 1946, the United Nations (UN) established the Population Division with the Department of Economic and Social Affairs (DESA). Over the years, the UN's Population Division has contributed to and supported the dialogue on population and development at the UN, producing regularly updated demographic estimates and projections for all countries as well as providing data that are essential for monitoring the status of implementation of internationally agreed development goals in the area of population, including those contained in the Sustainable Development Goals (SDGs).[2] Per the Population Division, the world population has more than tripled in size, rising from around 2.5 billion in 1950 to almost 7.9 billion in 2021. Projections suggest that the global population could grow to nearly 11 billion by the end of the 21st century when it is expected to stabilize (Fig. 1.6). The population growth that has occurred is the result of two trends: (1) widespread improvements in public health, nutrition, personal hygiene, and medicine, and (2) the elevated levels of fertility in many countries.

Despite these gains, significant disparities in life expectancy at birth remain across countries and regions, ranging from 50 years in some developing countries to more than 80 years in developed nations. Prominent levels of infant and child mortality persist in many regions, even though such deaths are largely preventable. The annual number of births worldwide is projected to remain around 140 million until 2050 and then fall steadily through 2100. Similarly, even though mortality rates are

---

[2] In 2016, the UN adopted 17 SDGs that become a call for action by all countries to promote prosperity while protecting the planet. Global health was represented in SDG-3: *Ensure healthy lives and promote well-being for all at all ages.*

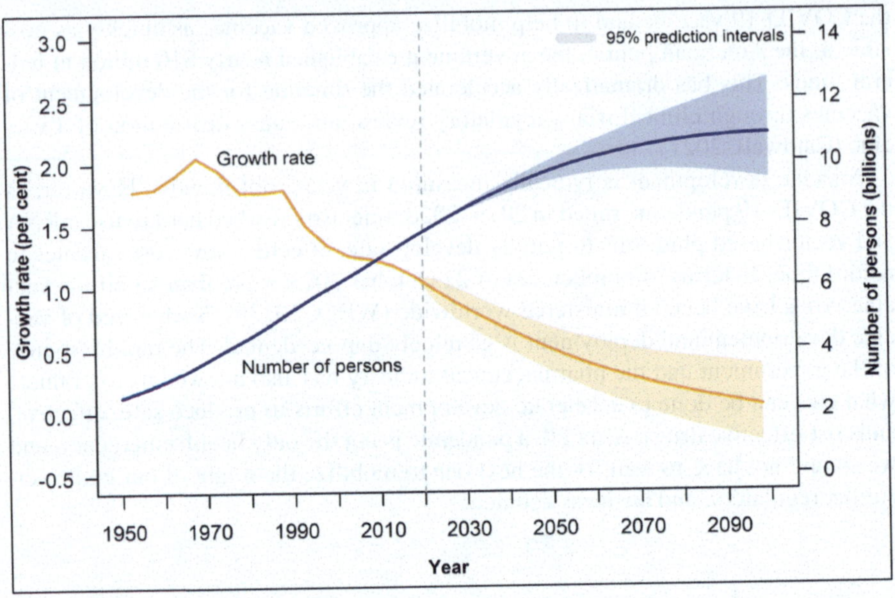

Fig. 1.6 Estimated global population size and annual growth rate (1950–2100) (DESA, 2021)

expected to continue falling over this period, the total number of deaths worldwide is projected to increase to nearly 93 million in 2050 due to population aging (Fig. 1.7), which shifts the population toward older ages where mortality risks are higher. The number of deaths worldwide is projected to approach 122 million by 2100 when it will be approaching the number of births (125 million) (DESA, 2021).

1.2.2 Leading Causes of Death

The past 60–80 years have revolutionized our knowledge of many medical conditions and associated diseases. We have developed medicines that are more specific, better tolerated, or target illnesses previously beyond our reach. However, the battle against disease is far from being won. In the twenty-first century, as our medical sciences and diagnostics advance, we face incredibly tough diseases such as HIV, Alzheimer's, and cancer, among many others. The Global Health Observatory (GHO) is a public health observatory established by the World Health Organization (WHO) to share data on global health, including statistics by country and information about specific diseases and health measures.

Causes of death and disability can be grouped into three large categories: (1) communicable (infectious diseases, along with maternal, perinatal, and nutritional conditions), (2) non-communicable (chronic diseases), and (3) injuries. Monitoring the yearly number of deaths helps to address their causes and adapt health systems

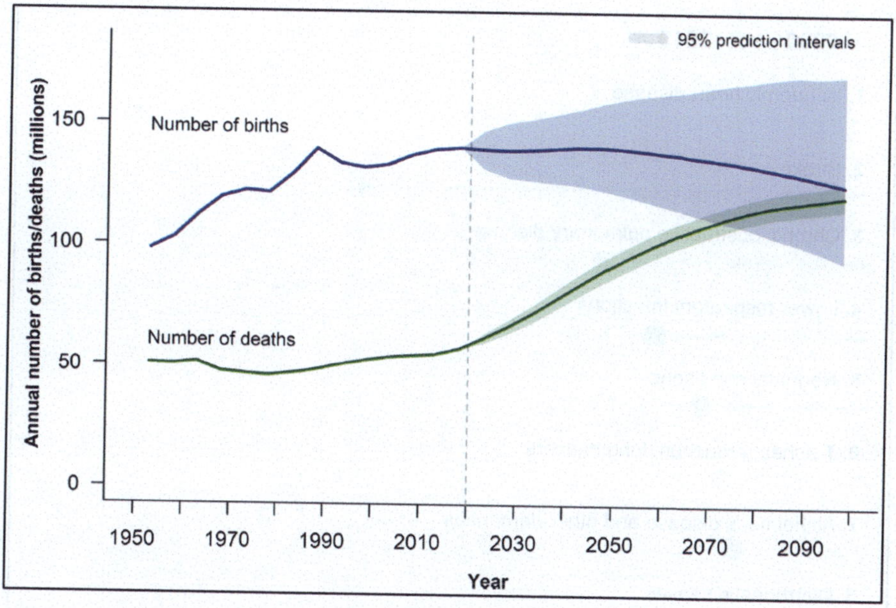

Fig. 1.7 Estimated global annual number of births and deaths (1950–2100) (DESA, 2021)

to react effectively, triggering responses in multiple sectors, from transportation (in addressing road traffic accidents) to food and agriculture (in tackling the rise in diabetes prevalence) and mental health support (in early detection of mental health conditions). Understanding why people die can help comprehend how people live to improve health services and reduce preventable deaths in every country, responding effectively to changing epidemiological circumstances. By providing key insights on mortality and morbidity trends, these estimates are a powerful tool to support informed decision making on health policy and resource allocation (WHO, 2023a).

As per the latest GHO factsheet, in 2019, the top 10 causes of death accounted for 55% of the 55.4 million deaths worldwide. Seven of the ten leading causes of death in 2019 were non-communicable diseases and accounted for 44% of all deaths or 80% of the top 10 (Fig. 1.8). The world's biggest killer is ischemic heart disease, responsible for 16% of the world's total deaths. Since 2000, this disease has had the most significant increase in deaths, rising by more than 2 million to 8.9 million deaths in 2019. Stroke and chronic obstructive pulmonary disease are the second and third leading causes of death, responsible for approximately 11% and 6% of total deaths, respectively. In general, deaths from non-communicable diseases are on the rise. There are, however, a few bright spots as well.

One of the most significant declines in the number of deaths is from diarrheal diseases, with deaths falling from 2.6 million in 2000 to 1.5 million in 2019. Other illnesses among the top ten causes of death in 2000 are no longer on the list. HIV/AIDS is one of them. Deaths from HIV/AIDS have fallen by 51% during the last 20 years, moving from the world's 8th leading cause of death in 2000 to the 19th in

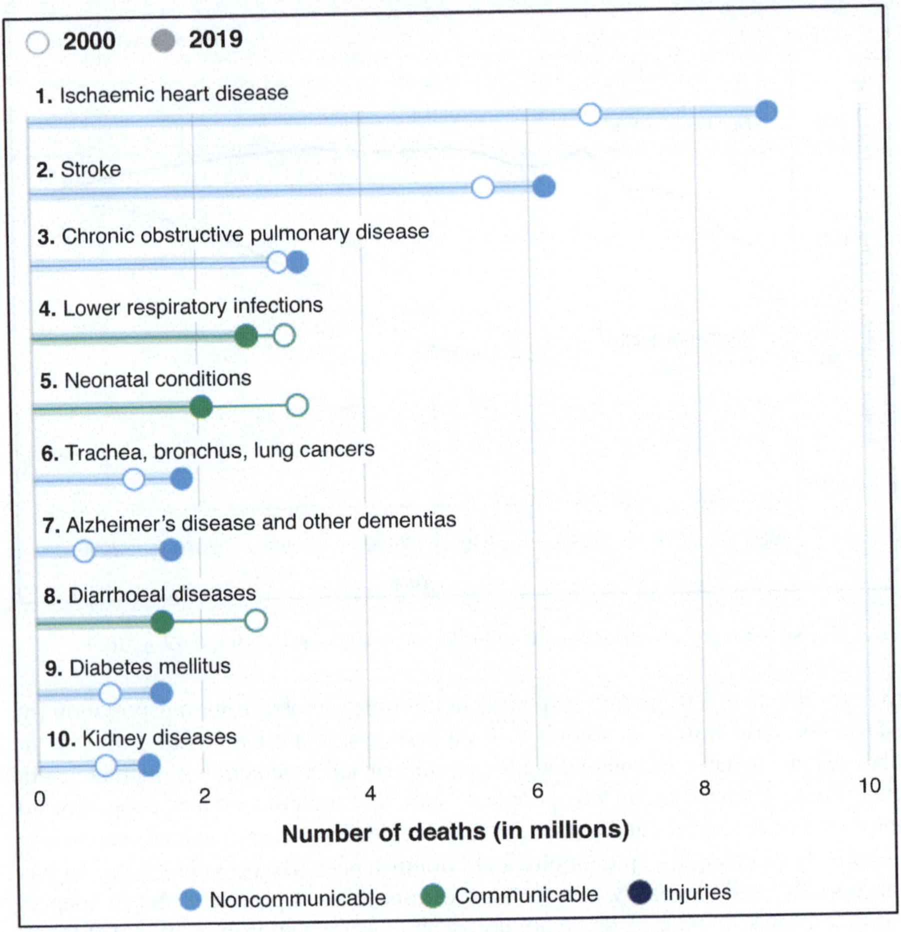

Fig. 1.8 Leading cause of death globally (WHO, 2023a)

2019 (WHO, 2023a). This is a testament to the multinational efforts to reduce the spread of HIV and concertedly work to identify the cause, make highly accurate testing available, undertake public health measures, accurate monitoring of disease, and develop effective medications. In particular, the effort to develop an effective treatment for HIV is remarkable in its speed and success. Clinical trials of antiretrovirals (ARVs) began in 1985—the same year that the first HIV test was approved—and the first ARV was approved for use in 1987. However, a single drug was found to have only short-term benefits. By 1995, ARVs were being prescribed in various combinations. A breakthrough in the HIV response was announced to the world at the 11th International AIDS Conference in Vancouver when the success of "highly active antiretroviral treatment" (HAART)—a combination of three ARVs—was reported, reducing AIDS-related deaths by between 60% and 80%. Unfortunately, not everybody would benefit from this life-saving innovation. Because of the high cost of ARVs, most low- and middle-income countries could not afford to provide treatment through their public programs. Such inequities generated outrage in communities and demands for affordable drugs and public treatment programs. Generic manufacturing of ARVs would only start in 2001, providing bulk, low-cost access to ARVs for highly affected countries, particularly in sub-Saharan Africa, where in 2000, HIV had become the leading cause of death (WHO, 2023a). Although much work still needs to be done (such as improving access to medications, increasing public awareness, and bridging the global health inequality gap), the positive outcomes achieved for curbing HIV provide hope and a template for the future. Therefore, it is increasingly vital for the medical community and pharmaceutical industry to innovate continuously to provide cures and relief to patients. Continued research in this area is a win-win scenario for society; however, from an economic perspective, a favorable outcome is not always guaranteed.

1.2.3 Economic Consequences of Disease

Philosophically, it can be postulated that health is a state of well-being on which everything else depends. Imagine how severely one's quality of life is impacted when they do not feel healthy emotionally, physically, or mentally. Not only is health fundamental to our existence, but its pursuit is also a lifelong endeavor. It also goes without saying that the healthcare industry is an integral part of any society. Healthcare is a highly regulated industry, and from an economic viewpoint, its operating model vastly differs from all other tangible commodities and assets, such as food, shelter, clothing, and other material possessions. The most obvious distinguishing characteristic of an individual's demand for medical services is that it is not steady in origin as, for example, for food or clothing, but irregular and unpredictable. Apart from preventive services, medical services afford satisfaction only in the event of illness, a departure from usual. Also, due to the specialized nature of medical diagnoses, consumers of medical services do not always have control over what they can purchase. The consumers must rely on the physician's professional advice. The customer cannot test the product before consuming it, and there is an element of trust in the relationship. Therefore, medical services have a specialized operating model that is significantly different from that of other typical commodities.

Measuring morbidity and mortality are vital considerations for estimating the burden of disease in populations. However, as can be expected, only focusing on morbidity and mortality effects provides an incomplete picture of the adverse impact of ill health on human welfare. In particular, the economic consequences of poor health can be substantial and range from unexpected increases in medical expenditure to complete economic ruin of an individual or family. Similarly, reduced functional capacity and lost income or productivity are often the primary risk factors for impoverishment. Poor levels of health may also adversely impact educational attainment and consequent levels of future income. At a societal level, poor population health is associated with lower savings rates, lower rates of return on capital, and lower levels of domestic and foreign investment; all these factors can and do contribute to reductions in economic growth. Increased health expenditures are also foreseen to impact the government's resources by reducing the amount of household taxable income and increasing the number of expenditures that must be covered out of public funds. The resulting impact on public budgets limits the government's ability to spend resources in other areas, particularly social expenditures such as education and public investments in infrastructure projects, sanitation, science and technology, and other strategic areas. Measurement of these various adverse impacts provides decision-makers with an indication of the extent to which a specific disease or, more generally, depleted health status disrupts or reduces economic production or consumption opportunities at the household or societal level (WHO, 2009).

Similarly, the positive association between health and wealth constitutes a compelling argument for more significant investment in health systems and services. Any healthcare intervention aims to improve health with available preventive measures, treatments, and medical procedures. The variation in health services provision across the country and increasing healthcare expenditure accentuate the need to utilize the sparse healthcare resources effectively. Similarly, at a global level, a country has a health system with better performance than another country if, for the same level of resources, it generates better health outcomes or if it generates the same outcomes but with fewer resources. Health systems are financed through taxes, like healthcare services owned by the state (national health services) or income-related social contributions (social security systems). In countries with high income per capita, the contributions to social security are essential and sustained, and to a great extent, so is the financing of the healthcare system (Jaba et al., 2014). Unfortunately, this is not valid globally; therefore, the quality of healthcare governance varies globally.

1.3 Pharmaceuticals: Fulfilling A Social Contract

Access to high-quality healthcare is a benchmark against a society's progress. Healthcare is also a costly enterprise whose expenditure only goes up with time! It is admirable to see that 75+ years after the formation of the UN and WHO, all countries have dedicated themselves to committing substantial resources toward healthcare. However, there remains a significant discrepancy in the number of resources dedicated to healthcare in different countries. One important healthcare metric is the percentage of a country's GDP spent on its citizens' healthcare, also known as Current Health Expenditure (CHE).[3] For example, the total US healthcare expenditure for 2019 was 16.8% of its GDP, totaling almost 3.8 trillion dollars, the highest in the world (CDC, 2022). In 2014, most countries spent 4.5–8.0% of their GDP on healthcare, with the world average at 9.9%. Data from selected countries over

[3] WHO maintains such databases as part of its Global Health Observatory.

Table 1.2 CHE as percentage of GDP of selected countries (WHO, 2024)

	2021	2011	2001
Afghanistan	21.83	8.56	Data Not Available
Argentina	9.71	9.42	8.37
Australia	10.54	8.54	7.68
Bangladesh	2.36	2.36	1.77
Belgium	11.04	10.36	8.15
Benin	2.58	3.00	3.30
Brazil	9.89	7.79	8.55
Canada	12.33	10.38	8.62
Chad	5.19	3.91	5.29
Chile	9.34	6.80	7.10
China	5.38	4.34	4.25
Colombia	9.02	6.78	5.94
Cuba	13.79	11.31	6.98
Denmark	10.82	10.41	8.45
Djibouti	2.88	3.38	3.20
Egypt	4.61	4.36	5.40
Ethiopia	3.21	4.47	4.72
France	12.31	11.18	9.71
Germany	12.93	10.78	9.92
Guatemala	6.90	5.96	6.07
India	3.28	3.25	4.26
Indonesia	3.71	2.99	2.12
Iran	5.77	6.10	5.16
Israel	7.90	6.84	7.02
Japan	10.82	10.49	7.24
Mali	4.47	4.01	4.52
Mexico	6.08	5.52	4.82
Netherlands	11.29	10.23	8.06
Nigeria	4.08	3.32	3.19
Peru	6.15	4.54	4.72
Russian Federation	7.39	4.79	5.24
South Africa	8.27	7.94	7.12
United Kingdom	12.36	9.92	7.34
United States of America	17.36	16.14	13.17
Zambia	6.62	3.96	4.83

20 years is shown in Table 1.2. Such benchmarking across various countries and through different periods allows governments to evaluate their progress against established healthcare goals.

Another metric is the Total Health Expenditure (THE), which is the sum of public and private health expenditures. It covers providing health services (preventive and curative), family planning activities, nutrition activities, and emergency aid designated for health but does not include water and sanitation. For example, in a 2016 global study, it was determined that the Healthcare Access and Quality (HAQ) Index, a healthcare metric, spanned from a high of 97.1 in Iceland, followed by 96.6 in Norway and 96.1 in the Netherlands, to values as low as 18.6 in the Central African Republic, 19.0 in Somalia, and 23.4 in Guinea-Bissau. However, discrepancies can also ensue in the local distribution of healthcare and could vary widely within a country. For example, in China, performance ranged from 91.5 in Beijing to 48.0 in Tibet. Similarly, India saw up to a 30-point disparity between its states (Fullman et al., 2018). This discrepancy highlights that access to safe and effective medicines remains disproportionate not only at a global level but also at a local level. Despite all well-intentioned efforts from stakeholders such as the government, patient advocacy groups, physicians, the pharmaceutical industry, and insurance companies, access to medicines remains a vexing issue.

1.3.1 Challenges with Access to Medication

The first synthetic pharmaceutical drug, Aspirin, was introduced in 1897. Since then, there have been incredible advances in drugs for a wide range of health concerns, including disease, mental health, and other conditions. Today, there are thousands of drugs on the market able to prevent, treat, and lessen the impact of ailments that would have been fatal just a few generations ago. Never in the past has our knowledge of science been so profound and the possibilities to treat all manner of diseases so great. Also, progress in fundamental research is nourishing an exceptional phase of developing medicines, vaccines, diagnostics, and medical devices.

Unfortunately, many people and communities needing effective prevention methods, life-enhancing and life-saving treatments, and rehabilitation do not receive them. Availability, affordability, and adaptation to specific settings and patient categories remain problematic in many regions and populations worldwide. Meanwhile, new threats in the form of new diseases are constantly emerging, for which global solutions must be found urgently (UN, 2016).

Responding to concerns raised by patients and health advocates worldwide, in 2015, UN Secretary-General Ban Ki-Moon convened a High-Level Panel on Innovation and Access to Health Technologies.[4] In outlining the mandate of this panel, the UN Secretary-General called upon the panelists to review and access proposals and recommend solutions for remedying the policy incoherence between the justifiable rights of inventors, international human rights law, trade rules, and public health in the context of health technologies. In its final report, the panel acknowledged that during the last 50 years, health outcomes have improved worldwide due to the improvement in medical innovation, nutrition, disease awareness, and sanitary conditions. Successful case studies include the development of the

[4]This panel was led by two former heads of state, Ruth Dreifuss of Switzerland and Festus Gontebanye Mogae of Botswana, together with 13 international experts with wide ranging perspectives.

polio vaccine, ARV therapy for Acquired Immune Deficiency Syndrome (AIDS), rotavirus vaccine, and hepatitis C medication. However, much work needs to be done, and despite many notable advancements, numerous challenges remain.

In some cases, progress has been uneven, leaving many people without access to the benefits of the advances. Causes for this vary, but in general, can be attributed to inadequate investments in R&D for diseases for which the market does not provide sufficient financial return. According to WHO, an estimated 1.7 billion people in 185 countries needed mass and/or individual treatment and care for neglected tropical diseases (NTDs) in 2014. Despite NTDs accounting for approximately 12% of the total disease burden, just 4% of therapeutic products registered between 2000 and 2011 were indicated for these diseases. Rapid point-of-care diagnostics, particularly for use in resource-limited settings and health emergencies such as Ebola and Zika, are essential, but since diagnostics can be complex and costly to develop, the net result is an inadequate number on the market.

Furthermore, despite progress, pediatric formulations remain scarce for conditions that affect children. Another potential crisis in waiting is bacterial antimicrobial resistance (AMR), which occurs when changes in bacteria cause the drugs used to treat infections to become less effective. AMR is exacerbated by the overuse of antibiotics—as well as poor waste management and climate change. It is estimated that if not controlled or reversed, AMR can cause up to 10 million deaths a year by 2050 and cost the global economy at least $100 trillion (UN, 2016).

Another challenge associated with healthcare is that cost is one of the key determinants of access, both when patients pay out-of-pocket and when the governments are making choices about what to fund under national health and treatment programs. According to the UN report, 400 million people worldwide lack healthcare, including access to medicines, vaccines, diagnostics, and medical devices. While most essential medicines (EM)[5] are off-patent, millions of people still do not have access to them due to regulatory inefficiencies, poor health education, unavailability of health insurance, and insufficient financial protection for those who must pay for some or all of their treatment. Often, the blame for this uneven access falls on the high costs of pharmaceuticals. This unidimensional thinking, unfortunately, belittles the innovative work done by pharmaceutical companies to bring medicine to market.

From a pharmaceutical company's perspective, substantial financial risk must be undertaken without any assurance of economic reward as pharmaceutical research is complicated, time-consuming, attritive, and costly, with estimated pre-approval

[5] Essential medicines are those that satisfy the priority health care needs of a population. They are selected with due regard to disease prevalence and public health relevance, evidence of efficacy and safety and comparative cost-effectiveness. They are intended to be available in functioning health systems at all times, in appropriate dosage forms, of assured quality and at prices individuals and health systems can afford. The WHO Model Lists of Essential Medicines are updated every 2 years by the Expert Committee on Selection and Use of Essential Medicines. The first Essential Medicines List was published in 1977, and the first Essential Medicines List for Children was published in 2007.

R&D costs of $2.6 billion per new drug (Schuhmacher et al., 2023). There is no doubt that given the costly nature of the drug development enterprise, access to medicines can be cost-prohibitive. Also, in the ever-changing business environment, pharmaceutical companies are under increased pressure to launch a new drug onto the market faster to achieve maximum market penetration and revenue in a limited time before the patent protection ends and generic competition begins. The successful launch of a new drug will pave the way for a pharmaceutical company's performance that enables R&D for new products in the future. Because of this business reality, pharmaceutical companies have seen a notable change in their operating model and footprint over the past few decades. Many had to undergo painful restructuring and workforce reductions because their traditional blockbuster model, i.e., finding effective medicines for common conditions and selling them in large volumes with substantial margins, is becoming extinct. Cheap generics are increasingly replacing top-selling drugs, and developing new drugs is more complicated and costlier because fewer opportunities exist, and R&D productivity has declined (Mattke et al., 2012; Mittal, 2019). Unfortunately, this hurts the sustainability of the healthcare system of which these products will be a part (Patel, 2017).

1.3.2 *Impact of Globalization on Intellectual Property Laws*

Access to essential drugs is part of the human right to health. Access to essential drugs depends on: (1) rational selection and use of medicines, (2) sustainable, adequate financing, (3) affordable prices, and (4) reliable health and supply systems. Since most poor people in developing countries currently pay for health care, including drugs, out of their own pockets, access to medicines is particularly sensitive to cost (WHO, 2001). To understand the high costs of pharmaceuticals, one must also appreciate how globalization and global trade agreements have impacted access to medicines.

The World Trade Organization (WTO) is an intergovernmental organization that regulates and facilitates international trade. With adequate cooperation in the UN system, governments use the WTO to establish, revise, and enforce the rules that govern international trade. The WTO is the only global international organization dealing with trade rules between nations. At its heart are the WTO agreements, negotiated and signed by most of the world's trading nations and ratified in their parliaments. The goal is to ensure that international trade flows smoothly, predictably, and freely. Its objectives are to promote: (1) non-discrimination, (2) progressive liberalization of trade barriers, (3) predictable policies and transparency, (4) competition, and (5) special provisions for developing countries (WHO, 2001). The WTO, headquartered in Geneva (Switzerland), is the world's largest international economic organization, with 164 member states representing over 98% of global trade and GDP (WTO, 2024).

Before the inception of the WTO, the General Agreement on Tariffs and Trade (GATT) functioned as a de facto international organization.[6] The GATT traces its origins to the 1944 Bretton Woods Conference, which laid the foundations for the post-World War II financial system and established two key institutions, the International Monetary Fund (IMF) and the World Bank (WB). The conference created rules governing exchange rates and international monetary cooperation, and those arrangements governed the conduct of world trade and payments. The arrangement provided for fixed exchange rates, with each member country fixing its currency value to the US dollar. The price of gold, fixed at $35 per ounce, applied to official transactions between national central banks. On 15 August 1971, the United States suspended the convertibility of the US dollar to gold, effectively bringing the Bretton Woods system to an end. With this, the world moved from a fixed to a flexible exchange-rate system (Hammes & Wills, 2005). Since then, gold prices have increased with time (Fig. 1.9).

A significant event that took place during the establishment of the WTO was a trade agreement known as the Trade-Related Aspects of Intellectual Property Rights (TRIPS), which required nations to provide minimum standard intellectual property (IP) protection for new technologies, including pharmaceutical products. IP provisions restrict the use and marketing of new drugs and grant exclusive rights to patent holders for 20 years to help offset the high costs required for R&D and incentivize investments in future clinical innovation with high uncertainty. That said, throughout the biomedical innovation cycle, IP rights may also hinder the possibility of inventor collaborations, innovation activities, production, and optimal global diffusion of affordable life-saving drugs (Mermelstein & Stevens, 2021).

TRIPS had a significant impact on public health policy. International conventions before TRIPS did not specify minimum standards for patents. Before the TRIPS agreement, patenting practices for pharmaceutical products varied widely and were less common in low- and middle-income countries (LMICs) that relied on access to generic drugs manufactured in emerging markets (Mermelstein & Stevens, 2021). Over 40 countries provided no patent protection for pharmaceuticals, many provided only process and not product patents, and the duration of patents was much less than 20 years in many countries (WHO, 2001). Shortly after TRIPS was introduced, concerns were raised by public stakeholders that the agreement would result in dramatic price increases of patented essential medicines that would disproportionally affect individuals and governments in less developed countries. According to the United Nations classification, the 46 least developed countries (LDCs) were exempted from granting or enforcing patent protection on medicines until 2033 (LDC Waiver). In 2005, a few amendments were made to the original TRIPS agreement to allow all other LMICs with little pharmaceutical manufacturing capacity to facilitate access to medicines by generic manufacturers through flexible provisions, namely compulsory licensing of medicines and government use of

[6] WTO officially commenced operations on 1 January 1995, pursuant to the 1994 Marrakesh Agreement, thus replacing the GATT. The WTO incorporates the principles of the GATT and provides a more enduring institutional framework for implementing and extending them.

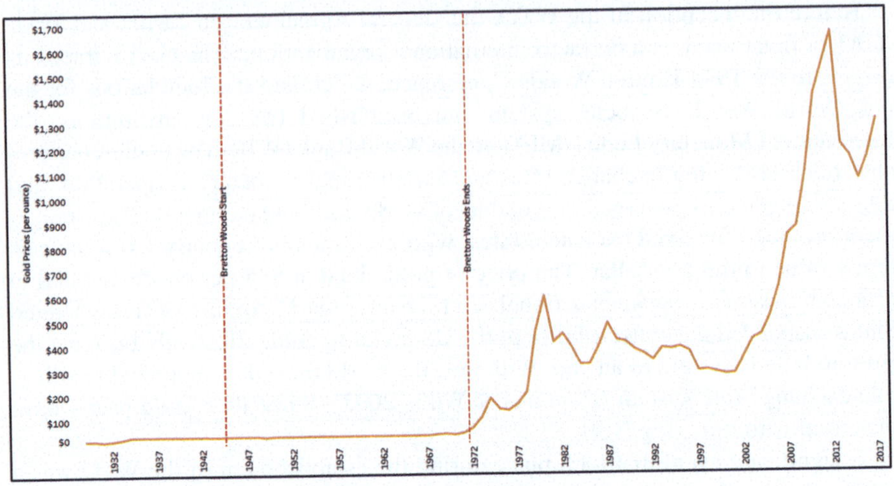

Fig. 1.9 Gold prices from 1929–2017 (Data Source: World Gold Council)

patents (Mermelstein & Stevens, 2021). A timeline of the global agreements and initiatives related to IP and access to medicines is shown in Fig. 1.10.

TRIPS allows member states to temporarily bypass patent holder's protections and issue compulsory licenses (CLs) assigned to government use or local private company to produce a patented product or compound for non-commercial domestic use or to export it to a developing country that lacks manufacturing capacity (Article 31bis). It can be granted on any ground, including public health emergencies, stock-outs, unaffordable prices, and patent holder's refusal to license. Researchers have collated the decision-making process of using CLs to improve access to medicines, as shown in Fig. 1.11 (Urias & Ramani, 2020). CLs are considered the key TRIPS-related flexibility for national governments in LMICs and played a central role in bringing down the price of patented medicines to treat HIV/AIDS and, more recently, Hepatitis C virus (HCV) infection by allowing mass production of generic substitutes (Mermelstein & Stevens, 2021).

1.3.3 *Improving the Efficiency of Pharmaceutical Development*

There is no doubt that pharmaceutical research and development is expensive. One of the biggest reasons for high expenses incurred during development is the potential for failures in clinical trials. For every 250 drugs that enter pre-clinical testing, only five will ever make it to human clinical trials, according to PhRMA. However, the trend in drug failures has changed during the past 20 years. The comparison of clinical attrition trends between 1991 and 2000 indicates that the principal component of drug failure in 1991 was the unacceptable pharmacokinetics profile in

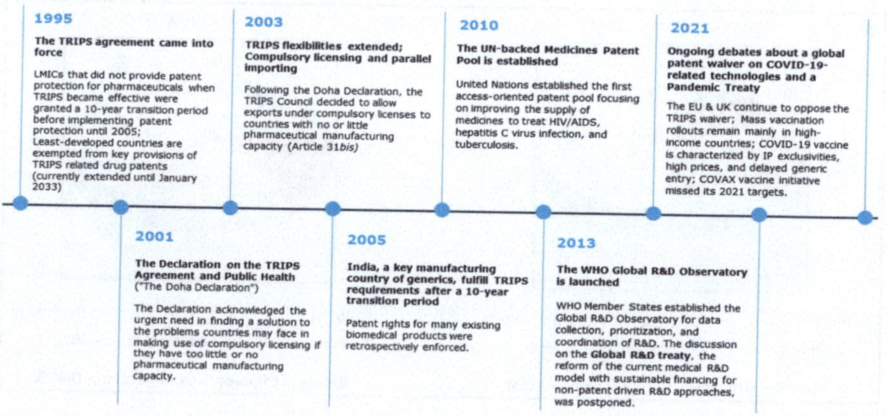

Fig. 1.10 Timeline of global agreements related to IP and access to medicines (Mermelstein & Stevens, 2021)

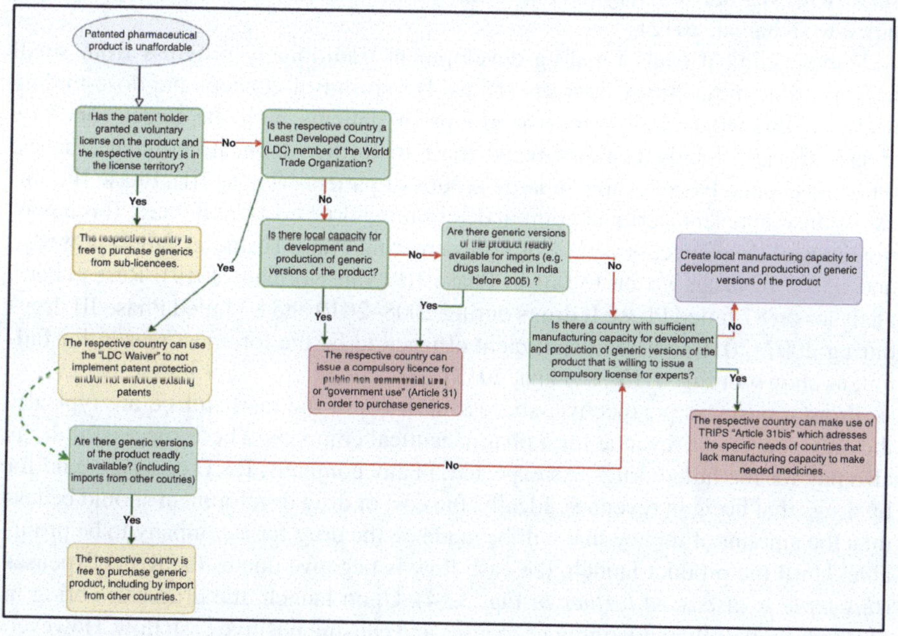

Fig. 1.11 Decision-making flowchart for usage of compulsory licenses (Urias & Ramani, 2020) (Reprinted with permission from Springer Nature)

humans. Since then, the industry has adopted several preclinical screens to address permeability, metabolism, distribution, and excretion issues, including allometric scaling for the projection of human pharmacokinetics profile. These efforts have paid off (Fig. 1.12), demonstrating marked improvement in success rates on drug metabolism and pharmacokinetics (DMPK) issues. This progress has continued

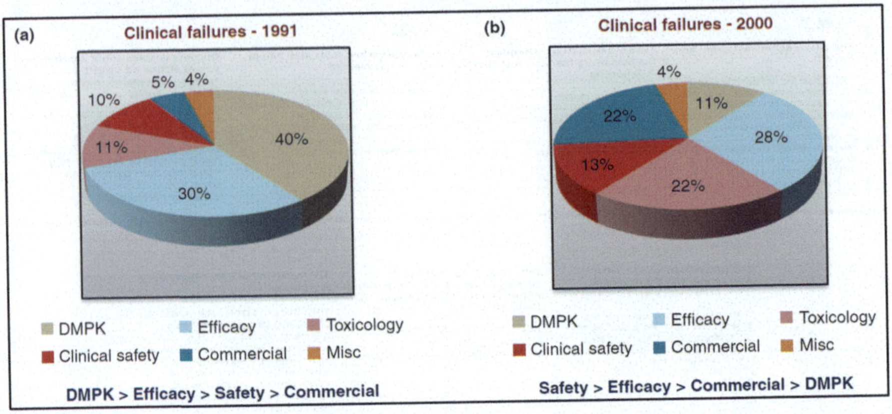

Fig. 1.12 Shifting paradigm in drug failures (Khanna, 2012) (Reprinted with permission from Elsevier)

since with regular delivery of improved DMPK candidates from the preclinical pipeline (Khanna, 2012).

Human clinical trials for drug development traditionally progress from small toxicity trials in healthy volunteers (Phase I) to proof-of-concept and dose-finding trials in somewhat more extensive groups of patients with the target condition (Phase II), and finally to randomized trials to further delineate clinical efficacy, outcomes, and adverse events in large groups of patients (Phase III). Phase II clinical studies represent a critical point in determining drug costs, and Phase II is a poor predictor of drug success: >30% of drugs entering Phase II studies fail to progress, and >58% of drugs go on to fail in Phase III (Van Norman, 2019). Researchers' analyses on 87 failed Phase II drugs during 2008–2010 and 83 failed Phase III drugs during 2007–2010 indicate insufficient efficacy to be the foremost reason for failure, as shown in Fig. 1.13 (Khanna, 2012).

From a business perspective, successful drugs in the marketplace are typically the only sources of revenue for a pharmaceutical company. These successful drugs also pay for the failed drugs. As expected, in any company, R&D costs are paid for by drugs that bring in revenues. Ideally, the cost of drug development should be less than the amount of money that will be made on the drug for a company to be profitable. Until the product launch, the cash flow is negative due to the R&D expenses (*this topic is discussed further in* Fig. 3.14). Upon launch, the drug sales bring in revenues, hopefully outweighing expenses and causing positive cash flow. However, there is an opportunity to reduce the development cost for new drug products. To stem the decline in productivity and to optimize its resources, it is increasingly necessary for the pharmaceutical industry to integrate business acumen and economics into rational product design. For example, numerous studies, such as excipient compatibility, prototype development, materials property evaluation, process development, tech transfer, and scale-up studies, are done as part of drug product

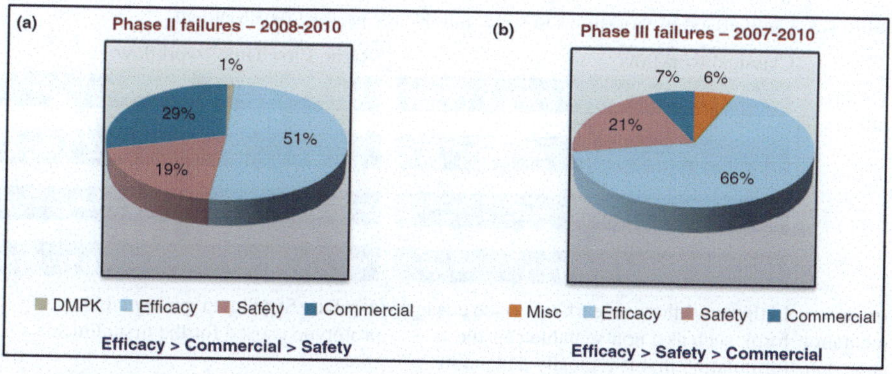

Fig. 1.13 Trends in late phase drug failures (Khanna, 2012) (Reprinted with permission from Elsevier)

development. In the case of solid oral dosages, this translates to working with biphasic materials that require an understanding of physical properties, mechanical properties, and machine-related variables. Each of these studies can be quite resource-intensive and, therefore, can add significant costs to the drug development process. However, efficiencies can be gained during development, such as integrating computer-aided design and manufacturing through strategic designs that conserve resources, time, and material. Similar approaches have been tried in other industries where successive improvements in components allow overall product costs to go down.

A desirable future state integrates fundamentals of economics in product design and development so that decision-making is parameterized, cost of goods can be lowered, wastage can be reduced, patient-centricity is built into the design, and manufacturing/distribution efficiencies can be gained. The financial benefits of such an approach could allow these savings to be passed on to the stakeholders and improve the value proposition of pharmaceuticals, which is critical to maintaining the innovation potential. Evidence of such a future is already here based on extensive research in areas such as reengineering pre-formulation and formulation studies through the "Right, First Time" concept that avoids the necessity to repeat expensive earlier clinical and toxicological studies by utilizing workflows that aim for market-ready dosage form in early clinical phases itself as shown in Table 1.3 (Leuenberger et al., 2013), improvements in manufacturing processes (e.g., in-built instrumentation, continuous manufacturing, adaptive manufacturing), and adoption of modeling & simulation in development. However, a more pervasive and critical discussion must enable strategic decision-making during product discovery, design, development, manufacturing, distribution, and end-consumer usage. Therefore, product development scientists should be more aware of the impact that the way that they plan their studies may have on the products' bottom line.

Table 1.3 Comparison of workflows in solid oral dosage product development

	Classical workflow	"Right, First Time" workflow
Visualization	Early Phase: Service dosage form (in general capsules) ↓ Change to tablets & Bioequivalence Testing ↓ Scale-Up Exercise ↓ Mass-production of final marketed form (*two-sigma* quality)	Market ready tablet dosage form (instead of service form) ↓ Small-scale Production ↓ Scale-Up Exercise (computer-assisted) ↓ Mass-production of final marketed form (*six-sigma* quality)
Implementation approach	In this workflow, an earlier service dosage form, such as a non-scalable capsule formulation that is typically a blend in capsule (or a drug in capsule), is utilized to initiate Phase I clinical trials.	A robust Six Sigma quality tablet prototype is used for the first clinical trials.
	The simple, not-optimized capsule formulation is often replaced later by a tablet as the final marketed dosage form.	All collected data since Clinical Phase I can be used to further prototype optimization. No bioequivalence test is needed.
	The bridging from capsule to tablet may require additional toxicological and bridging PK studies.	The stability data of the prototype formulation being close to the final marketed dosage form can be used for registration purposes.
	Stability data has to be regenerated for each new clinical formulation	

Adapted from Leuenberger et al. (2013) and Leuenberger et al. (2016) (reprinted with permission from Elsevier)

Could concepts such as 'Right, First Time' and similar workflows that are used to improve product quality also be proactively used to lower cost of goods (COGs) and pass the savings onto stakeholders?

1.3.4 Aligning Pharmaceutical Development to Sustainable Health Systems

In a classical health economics model, everyone is viewed as both a producer and a consumer of health. It assumes that individuals inherit an initial "health-stock" that depreciates over a lifetime—at an increasing rate, at least after some stage—and can be increased by investment. Direct inputs into the production of gross investments in the health-stock include time, medical care, diet, exercise, housing, and other market goods (Grossman, 1972). Another critical aspect of the health economics model is the availability of "inducers" that can alter the production process of the health-stock. This is where healthcare systems come in. Researchers measure the success of health systems using three attributes: (1) healthy people, (2) superior care, and (3) availability of treatments. Numerous metrics have been designed to measure these attributes, ranging from evidence-based healthcare administration, to enhanced access to Medicare/Medicaid, and other socio-political initiatives. However, the sustainability of these health systems is under constant scrutiny. A sustainable health system has three key attributes: (1) affordability for all stakeholders, (2) acceptability to patients and health professionals, and (3) adaptability to emerging research and challenges (Fineberg, 2012). Significant discussions in the literature exist on the critical evaluation of health systems and their socio-economic-political impact. One key piece that can also help improve the sustainability of health systems is the strategic and efficient development of drug products.

The development of drug products is complex and challenging but a worthy undertaking for the betterment and advancement of civilization. In our current times, the science of drug development is an established field that requires a dedicated and synergistic partnership between various subject matter experts. Given the experience that we now have in drug development, the steps taken toward establishing a drug's safety, efficacy, and the process for its commercialization have long been standardized. The whole pharmaceutical (and biopharmaceutical) community is exceptionally committed to commercializing new medicines despite its challenges. Given its innovative approaches to discovering cures for unmet medical needs, the pharmaceutical industry has always emphasized R&D, delivering one of the highest ratios of R&D investment to net sales compared with other industrial sectors.

More than 750 companies are listed as either biotechnology or major pharmaceutical companies on the NYSE and NASDAQ combined (Table 1.4). These companies cover a spectrum of business operating models, from fully integrated research companies with thousands of employees to virtual entities with minimal permanent staff. No matter their *modus operandi*, all these companies are developing new drug products for many therapeutic indications. From a societal viewpoint, with the number of resources spent on each new drug candidate, the total costs start to add up quickly! This further necessitates that any efforts that can help to lower the development time and costs should be carefully reviewed and, if possible, integrated into the product development strategy. Pharmaceutical innovation without integrating economics into product design is an inherently futile and non-sustainable proposition. Therefore, the time to discuss how the development of robust and productive healthcare options can be achieved economically has now come, for, after all, no industry is immune to the laws of economics. With this premise, it is critical to spur a conversation that aims to integrate drug development, economics, and efficient manufacturing in a fashion that improves the accessibility of drugs to patients. Such conversations allow drug development scientists to get closer to building an "empathetic bridge" and be more aware of the patient's needs.

Table 1.4 Industry profile for the healthcare sector of the US stock market

Industry name	# of individual companies	Total market cap (US $) (as of March 02, 2024)
Biotechnology	692	1487.48B
Medical devices	150	891.15B
Drug manufacturers—specialty & generic	100	305.57B
Diagnostics & research	63	770.80B
Medical instruments & supplies	58	455.75B
Health information services	56	137.22B
Medical care facilities	49	195.26B
Drug manufacturers—general	17	2920.70B
Healthcare plans	12	871.39B
Pharmaceutical retailers	10	19.29B
Medical distribution	9	153.52B

Adapted from (Stock_Analysis (2024))

1.4 Summary—Build Our Collective Future Together

The innate desire to bridge the health inequity gap is not new. However, the virtues of patience and maturity to understand someone else's perspective without dissolving a worthy discussion into a rancorous and unproductive debate are in short supply in recent times. In this debate, sadly, more often than not, the pharmaceutical industry is constantly under scrutiny for good and bad reasons alike. Pharmaceutical scientists must realize that no matter how divisive the debate becomes, there is no question that the development of drug products for consumption is complex and challenging but a worthy undertaking for the betterment and advancement of civilization. When an innovative pharmaceutical company launches a new product, it is a cause for celebration as the new drug could help patients suffering from rare disorders to common diseases—to gain hope for improved quality of life and, in some cases, enhanced chances of surviving life-threatening illnesses. Drug development is too noble a cause not to give it all we have as scientists! We, however, can learn techniques and design philosophies to integrate the laws of economics into our products efficiently. Not only will this help reduce the costs of goods, but it will also improve the accessibility of medicines for our patients. After all, society always wins with every new drug discovered and commercialized. ***Can any other industry make such a bold claim?***

References

ASQ. (2024). *What is benchmarking?* Retrieved January 1, 2024 from https://asq.org/quality-resources/benchmarking

Caldwell, A. (2021). *How were researchers able to develop COVID-19 vaccines so quickly?* University of Chicago Office of Communications. Retrieved February 11, 2023 from https://news.uchicago.edu/story/how-were-researchers-able-develop-covid-19-vaccines-so-quickly

CBO. (2021). *Research and development in the pharmaceutical industry*. Retrieved from https://www.cbo.gov/publication/57126
CDC. (2022). *Health care expenditures*. CDC. Retrieved October 6, 2022 from https://www.cdc.gov/nchs/hus/topics/health-care-expenditures.htm#:~:text=In%202019%2C%20current%20%28nominal%20dollars%29%20spending%20for%20national,dollars%20removes%20the%20effect%20of%20inflation%20on%20spending
De Vries, J. (1994). The industrial revolution and the industrious revolution. *The Journal of Economic History, 54*(2), 249–270. http://www.jstor.org/stable/2123912
DESA, P. D. (2021). *Global population growth and sustainable development*. (UN DESA/POP/2021/TR/NO. 2). United Nations Retrieved from https://www.un.org/development/desa/pd/sites/www.un.org.development.desa.pd/files/undesa_pd_2022_global_population_growth.pdf
Fineberg, H. V. (2012). A successful and sustainable health system—How to get there from here. *New England Journal of Medicine, 366*(11), 1020–1027. https://doi.org/10.1056/NEJMsa1114777
Fullman, N., Yearwood, J., Abay, S. M., Abbafati, C., Abd-Allah, F., Abdela, J., Abdelalim, A., Abebe, Z., Abebo, T. A., Aboyans, V., Abraha, H. N., Abreu, D. M. X., Abu-Raddad, L. J., Adane, A. A., Adedoyin, R. A., Adetokunboh, O., Adhikari, T. B., Afarideh, M., Afshin, A., et al. (2018). Measuring performance on the Healthcare Access and Quality Index for 195 countries and territories and selected subnational locations: A systematic analysis from the Global Burden of Disease Study 2016. *The Lancet, 391*(10136), 2236–2271. https://doi.org/10.1016/S0140-6736(18)30994-2
GALE. (2022). *Industrial revolution*. https://www.gale.com/open-access/industrial-revolution
Grossman, M. (1972). *The demand for health: A theoretical and empirical investigation*. National Bureau of Economic Research (NBER).
Hammes, D., & Wills, D. (2005). Black gold: The end of bretton woods and the oil-price shocks of the 1970s. *The Independent Review, 9*(4), 501–511. http://www.jstor.org/stable/24562081
IPCC. (2023). *Climate change 2023: Synthesis report* (Contribution of Working Groups I, II and III to the Sixth Assessment Report of the Intergovernmental Panel on Climate Change, Issue. IPCC. https://www.ipcc.ch/report/ar6/syr/
Jaba, E., Balan, C. B., & Robu, I.-B. (2014). The relationship between life expectancy at birth and health expenditures estimated by a cross-country and time-series analysis. *Procedia Economics and Finance, 15*(14), 108–114.
Jarus, O. (2022). *The worst epidemics and pandemics in history*. Future US Retrieved January 14, 2023 from https://www.livescience.com/worst-epidemics-and pandemics-in-history.html
Khanna, I. (2012). Drug discovery in pharmaceutical industry: productivity challenges and trends. *Drug Discovery Today, 17*(19), 1088–1102. https://doi.org/10.1016/j.drudis.2012.05.007
Leuenberger, H., Puchkov, M., Schneider, B., & Baselland, I. (2013). Right, first time concept and workflow a paradigm shift for a smart & lean six-sigma development. *Swiss Pharma, 35*(3).
Leuenberger, H., & Leuenberger, M. N. (2016). Impact of the digital revolution on the future of pharmaceutical formulation science. *European Journal of Pharmaceutical Sciences, 87*, 100–111.
Mattke, S., Klautzer, L., & Mengistu, T. (2012). Medicines as a service: A new commercial model for big pharma in the postblockbuster world. In *Medicines as a service* (pp. 1–12). RAND Corporation. http://www.jstor.org/stable/10.7249/op381hlth.1
Mermelstein, S., & Stevens, H. (2021). TRIPS to where? A narrative review of the empirical literature on intellectual property licensing models to promote global diffusion of essential medicines. *Pharmaceutics, 14*(1). https://doi.org/10.3390/pharmaceutics14010048
Mittal, B. (2019). Chapter 3—Pharmaceutical productivity: Challenges and opportunities. In B. Mittal (Ed.), *How to integrate Quality by Efficient Design (QbED) in product development* (pp. 59–86). Academic. https://doi.org/10.1016/B978-0-12-816813-4.00003-9
Ostwald, D., Cramer, M., Albu, N., & Tesch, J. (2020). *The global economic impact of the pharmaceutical industry*. https://www.wifor.com/uploads/2021/06/WifOR_Global_Economic_Footprint_Study_September_2020.pdf

Patel, N. A. (2017). Fee-for-value in the pharmaceutical industry: A policy framework applying data science to negotiate drug prices. *Journal of Law and the Biosciences, 4*(1), 205–215. https://doi.org/10.1093/jlb/lsx002

PhRMA. (2017). The biopharmaceutical industry's role in fueling the U.S. *Economy and Global Competitiveness*. https://phrma.org/-/media/Project/PhRMA/PhRMA-Org/PhRMA-Org/PDF/Industry-Profile-2021/The-Biopharmaceutical-Industrys-Role-in-Fueling-the-US-Economy-and-Global-Competitiveness.pdf

Schuhmacher, A., Hinder, M., von Stegmann und Stein, A., Hartl, D., & Gassmann, O. (2023). Analysis of pharma R&D productivity—A new perspective needed. *Drug Discovery Today, 28*(10), 103726. https://doi.org/10.1016/j.drudis.2023.103726

Stock_Analysis. (2024). *Stock sectors & industries* https://stockanalysis.com/stocks/industry/

Tichy, E. M., Hoffman, J. M., Suda, K. J., Rim, M. H., Tadrous, M., Cuellar, S., Clark, J. S., Ward, J., & Schumock, G. T. (2022). National trends in prescription drug expenditures and projections for 2022. *American Journal of Health-System Pharmacy, 79*(14), 1158–1172. https://doi.org/10.1093/ajhp/zxac102

UN. (2016). *The United Nations Secretary-General's high-level panel on access to medicines report—Promoting innovation and access to health technologies*. UN. https://www.unsgaccessmeds.org/final-report

Urias, E., & Ramani, S. V. (2020). Access to medicines after TRIPS: Is compulsory licensing an effective mechanism to lower drug prices? A review of the existing evidence. *Journal of International Business Policy, 3*(4), 36–384. https://doi.org/10.1057/s42214-020-00068-4

Van Norman, G. A. (2019). Phase II trials in drug development and adaptive trial design. *JACC: Basic to Translational Science, 4*(3), 428–437. https://doi.org/10.1016/j.jacbts.2019.02.005

Weger, U. W., Wagemann, J., & Tewes, C. (2019). Editorial: The challenges and opportunities of introspection in psychology: Theory and method. *Frontiers in Psychology, 10*, 2196. https://doi.org/10.3389/fpsyg.2019.02196

WHO. (2001). *Globalization, TRIPS and access to pharmaceuticals.*

WHO. (2009). *WHO guide to identifying the economic consequences of disease and injury.* Retrieved from https://www.who.int/choice/publications/d_economic_impact_guide.pdf

WHO. (2023a). *Global health estimates: Leading causes of death.* WHO. Retrieved December 10, 2023 from https://www.who.int/data/gho/data/themes/mortality-and-global-health-estimates/ghe-leading-causes-of-death#:~:text=Ischaemic%20heart%20disease%20was%20the,over%20the%20last%20two%20decades

WHO. (2023b). *WHO coronavirus (COVID-19) dashboard.* WHO. Retrieved from https://covid19.who.int/info

WHO. (2024). *Current health expenditure (CHE) as percentage of gross domestic product (GDP) (%).* https://www.who.int/data/gho/data/indicators/indicator-details/GHO/current-health-expenditure-(che)-as-percentage-of-gross-domestic-product-(gdp)-(-)

WTO. (2024). *The WTO*. Retrieved January 18, 2024 from https://www.wto.org/english/thewto_e/thewto_e.htm

Chapter 2
Sustainable Global Health Systems: Our Collective Responsibility

One person can make a difference, and everyone should try.

John F. Kennedy

Contents

Abstract The root causes of health inequities are complex, arising from the interaction of various political, economic, and social factors. The usual suspects are infrastructure issues, lack of education/awareness, conflicts/wars, poverty, and limited access to adequate healthcare. One possible solution is responsible governance, which requires data monitoring and trend analysis to allocate resources efficiently. Another approach most discussed to bridge the health inequality gap is that of Universal Health Coverage (UHC). UHC will allow all people access to the medical services they need, when and where they need them, without financial hardship or catastrophic health spending. Many countries at varying stages of economic development have shown that it is possible to make substantial progress towards UHC. Nevertheless, the world still has a long way to go. The WHO estimates that half of the world's population cannot access the healthcare they need. Over 930 million people spend at least 10% of their household income on healthcare, and

B. Mittal, *Sustainable Global Health Systems and Pharmaceutical Development*, AAPS Advances in the Pharmaceutical Sciences Series 60, https://doi.org/10.1007/978-3-031-50415-0_2

100 million people are driven into poverty each year. Fortunately, these numbers can be lowered through improved accessibility and availability of medicines and other healthcare resources. A 2010 report from WHO estimates that 20% to 40% of all health spending is wasted through inefficiency and points to 10 specific areas where better policies and practices could increase the impact of expenditures, sometimes dramatically. Spending on medicines is one area where efficiencies can be gained, and wastage can be reduced. Various organizations are creating practical tools to measure the global use of medicines. Investing resources more wisely can help countries move closer to universal coverage without increasing spending.

Keywords Global health governance · Sustainable development goals · Universal health coverage · Service coverage index

2.1 Introduction

Health is the product and reflection of society's attention to an adequate standard available to all in the conditions in which its population lives (Friel, 2017). The development of a society can be judged by the quality of its population's health, by how fairly health is distributed across the social spectrum, and by the degree of protection provided from a disadvantage because of ill health. Unfortunately, there is a wide distribution of life expectancy at birth in various parts of the world. It is a sad reality that a child born today can expect to live less than 50 years in some parts of the world, while in other parts, a child can expect to live for more than 80 years. Issues such as child malnutrition, maternal mortality, the prevalence of HIV/AIDS, and many others continue to remain a challenge of our times!

To provide cures and relief to patients, it is increasingly vital for the medical community and pharmaceutical industry to innovate continuously. However, it should never be forgotten that the medicines and treatments we take for granted today were the results of laborious and meticulous scientists, doctors, and healthcare providers who paved the way for the betterment of society. It is also important to realize that a discovery was often made by accident only because the observer was astute in connecting the dots and had the mental fortitude to see the discovery through the scientific method. Serendipity in research is often finding the unexpected, which, when recognized, happily changes the course of the research for good. Often, such serendipitous findings were made in the most unlikely of scenarios. One such story is that of Florence Nightingale and her contributions to healthcare data management.

2.1.1 *Philosophical Digression: A Humanitarian and Statistician*

Florence Nightingale (1820–1910) is one of the most famous and easily recognizable Victorians. In 2002, Nightingale was ranked #52 in BBC 's list of the 100 GreatestBritons following a UK-wide vote. In 2006, the Japanese public ranked Nightingale #17 in The Top 100 Historical Persons. Several churches in the Anglican Communion commemorate Nightingale with a feast day on their liturgical calendars. Although she is known to everyone as the "Lady with the Lamp" who established the modern nursing profession by providing proper training facilities, we know less about her role as the "Passionate Statistician" (Magnello, 2010).

Light years ahead of their time, Nightingale's parents aimed to improve social conditions for everyone and approved of women's education. Florence's father, William, a Trinity College graduate, included Greek, Latin, and mathematics in her lessons. By the time Florence was 9 years old, she was already organizing statistical data and collaborating on a project with male contemporaries. As a young woman, she learned about statistics from the Victorian epidemiologist William Farr (1807–1883) and the Belgian astronomer and social statistician Lambert Adolphe Jacques Quetelet (1796–1874). By 1853, Florence became Superintendent of a small hospital in London. In her new position, she reorganized the bookkeeping and established a more efficient infrastructure within the hospital. Later that year, she began distributing health administration questionnaires to different hospitals in Europe and compiled statistical data (Magnello, 2010).

Florence Nightingale's most famous contribution came during the Crimean War (1853–1856), which became her central focus when reports got back to Britain about the horrific conditions for the wounded. In 1854, she and the staff of 38 volunteer nurses that she had trained were sent to the Ottoman Empire. Her team found that poor care for wounded soldiers was being delivered by overworked medical staff in the face of official indifference. Medicines were in short supply, hygiene was being neglected, and mass infections were common, many fatal. There was no equipment to process food for the patients. Once Nightingale and her team took charge of the situation, the death rates reduced significantly from 42% to 2% by improving hygiene or calling for the Sanitary Commission. During the first 7 months of the Crimean campaign, Nightingale collected statistical data and concluded that more soldiers died in the barracks than on the battlefields. Nightingale understood the power of statistics and used them to corroborate her arguments (Magnello, 2010).

Armed with her statistical findings, upon returning to London in August 1856, she met William Farr, then Statistical Superintendent of the General Record Office (now the Office for National Statistics). They analyzed the Crimean data and used graphs to show their findings to various government officials. Nightingale developed a flair for devising graphic methods and created a very impactful Polar Area

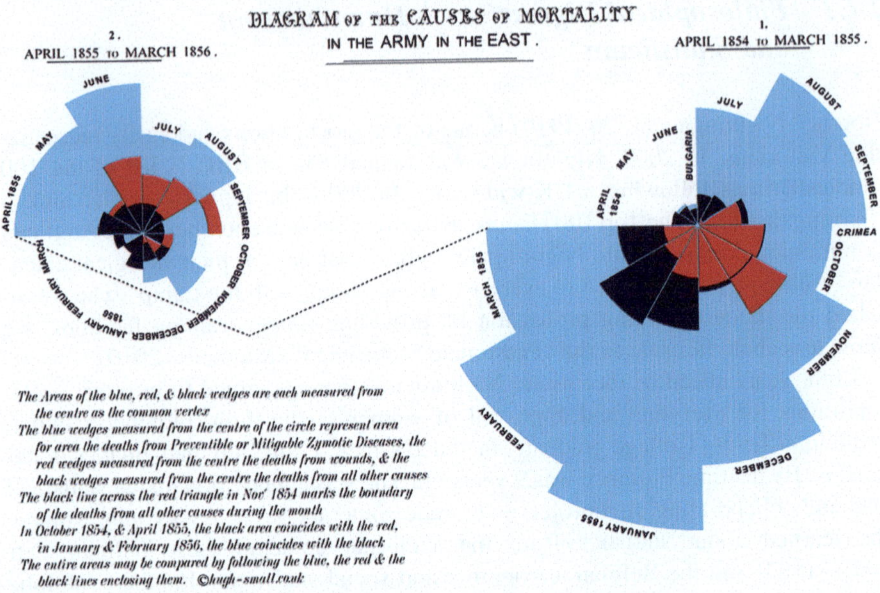

Fig. 2.1 Florence Nightingale's polar area graph. (Friendly, 2009)

Graph[1] (Fig. 2.1). The graph was cut into 12 equal angles, representing the year's months. Each wedge represents the number of deaths from the disease. There were no fatalities at the start of the war in April 1854, but as the months passed, the numbers increased, and the radii of the wedges got bigger until January 1855. She explained that this was because, as the fighting went on, the hospitals became more crowded and dirtier, so there was more disease. Her graph not only dramatized the extent of the needless deaths amongst the soldiers during the Crimean War, but it was used as a corrective tool to persuade government officials and the medical profession that deaths were preventable if sanitation reforms were implemented in hospitals (Magnello, 2010). Nightingale was not the inventor of polar area diagrams, but rather, she was arguably the first to use this and other statistical graphs for political persuasion and widespread impact (Friendly, 2009).

Although Florence Nightingale is highly revered for reforming nursing in the nineteenth century, she deserves more recognition for revolutionizing nursing by implementing medical statistics. Her analytical work led to a decline in preventable hospital deaths and enabled her to introduce mathematical rigor into nursing and medical research. Moreover, in 1858 she was the first woman to be elected a Fellow

[1] In a polar area chart, the circle is divided into angles or "wedges" of the same size for each category. The radius of each wedge is then equal to the square root of the count (or frequency) for that category. The square root is used for the radius because the area of a circle is πr^2, and using the square root of the frequency for radius means that the wedges in the polar area chart are still proportional to one another, like in the pie chart.

of the Statistical Society of London (now the Royal Statistical Society). In 1860, she established the Nightingale Training School for Nurses at St. Thomas in London. In 1912, the International Committee of the Red Cross instituted the Florence Nightingale Medal, awarded every two years to nurses or nursing aides for outstanding service. Since 1965, International Nurses Day has been celebrated on her birthday (May 12) (Magnello, 2010).

2.1.2 Opportunity for Pharmaceutical Scientists: Aspire for Global Health Solutions

Technology is moving at a breakneck speed. The world is also changing as the once-abundant natural resources are depleting at a much faster pace. In the meantime, practical solutions for our changing operating environment are also being developed. The development of these solutions, however, may not always be in the field of pharmaceutical development. Therefore, an awareness of what is going on in other industries helps us pharmaceutical scientists to appreciate, understand, and evaluate what it is that we can implement in our field to make our resource utilization more optimal and cost-effective. We may not realize during our daily work that our society is more connected today than ever in the past. Not only do we share common problems such as overpopulation, massive land-use changes, pollution, climate change, dwindling resources, and food security, but we also must collaborate to achieve collective solutions for the benefit of our future generations. Therefore, one should always strive to be "intellectually curious".

One of the biggest common problems we share is global health and its associated inequality. Global health is an organizing framework through which the effects of identity, social position, policies, institutional practices, and geography of multiple populations of people intersect with the health of animals and our changing environments. This framework was codified into 12 principles known as the 2004 Manhattan Principles. The Manhattan Principles were developed at the Wildlife Conservation Society's meeting on 'One World, One Health: Building Interdisciplinary Bridges to Health in a Globalized World' that recognize that human, animal, and environmental health are not mutually exclusive and that each is shaped by the relationships between them. The principles recognized that decision-making processes are critical to maintaining the integrity of biodiversity, food supplies, and economies. They stressed the impact of decisions on relationships between ecosystem resilience and disease emergence and spread patterns. The principles embedded global disease prevention, surveillance, monitoring, control, and mitigation in biodiversity conservation work and called for holistic, integrative, and ethical approaches to minimize social inequity. They acknowledged the overlapping agendas linking human, environmental, and animal health and called on the global community to unite on security (Jeggo et al., 2019).

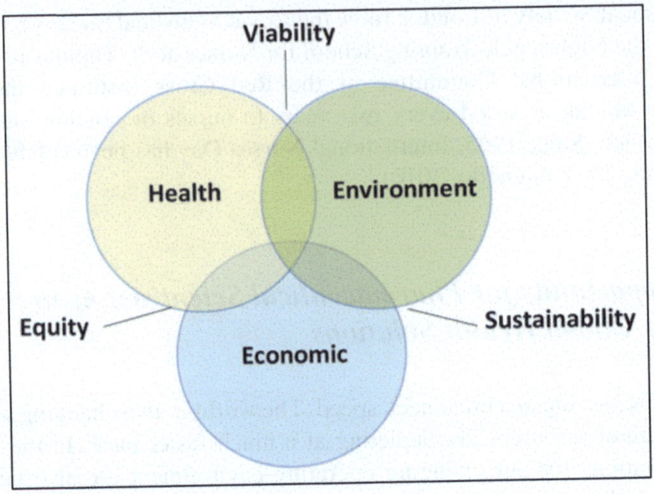

Fig. 2.2 Model of human development. (Initially proposed by Hancock (1993), reproduced from Kjærgård et al. (2013)) (reprinted with permission from Oxford University Press)

Other researchers have posited that the concept of health should be understood in a broader context than a narrower viewpoint, which considers it as just the absence of disease. Health is part of the social dynamics of social organization lifestyles and is influenced by the bio-physical environment. Hence, health is determined by the complex social and economic system context, the bio-physical environment, and the person's characteristics and behaviors. Health promotion implies a paradigm shift from comprehending health as the absence of disease (the bio-medical approach) to a socio-ecological understanding of health focusing on strength, resilience, and assets to health (Kjærgård et al., 2013). A seminal article written in 1993 proposed a model of human development (Fig. 2.2) that enabled diverse groups and sectors, such as health, human rights, education, environment, economy, etc., to coalesce around a common objective (Hancock, 1993). The model shows the links between health or social well-being environmental and economic well-being with a focus on equity and sustainability (Kjærgård et al., 2013). Therefore, any economic activity that aims to help with global health equity must be rooted in sustainability-driven initiatives that strive to meet the needs of the present without compromising the ability of future generations to meet theirs.

What are some of the latest developments in sciences and manufacturing technologies that we could apply in sustainable pharmaceutical product development?

2.2 Global Health Equity

Access to good healthcare is one of the fundamental human rights. Health equity gives everyone a fair chance to lead a healthy lifestyle irrespective of race, ethnicity, or socioeconomic standing. Health equity is achieved when it creates fair opportunities and eliminates the gaps in health outcomes between different social groups. Global health is created to improve health worldwide, reduce disparities, and protect against threats that disregard national borders (Brodit & Radwan, 2018). Better health equity can help drive economic growth, bringing peace and prosperity. Researchers have studied the link between health expenditures and the population's health status, which clearly supports the idea that policies should be oriented toward reducing health inequalities among world countries (Jaba et al., 2014). Unfortunately, such a desirable state is not easy to attain as significant barriers exist.

The root causes of health inequities are complex, arising from the interaction of various political, economic, and social factors. The usual suspects are infrastructure issues, lack of education/awareness, conflicts/wars, poverty, and inadequate access to healthcare. The financial sustainability of publicly funded healthcare systems is a challenge to policymakers in many countries as healthcare absorbs an ever-increasing share of national wealth and government spending. New technology, populations, and rising public expectations of the healthcare system are often cited as reasons why these systems need ever-increasing funding and why universal and comprehensive public systems are unsustainable (Birch et al., 2015). However, there is cause for hope here. While health inequality can be defined as the difference in health between different social groups and nations, health inequity is that part of the difference that could be avoided or remedied (Friel, 2017). One way to interpret this is by realizing that reducing health inequity is a moral responsibility and a societal obligation that beckons us all to work toward reasonable equality among people in health outcomes. We all have a role to play in striving to achieve health equity.

In 2005, the WHO set up the Commission on Social Determinants of Health (CSDH). CSDH was mandated to collect, collate, and synthesize global evidence on the social determinants of health, their impact on inequality, and to make recommendations for action. In its landmark report issued in 2008, the CSDH acknowledged that the causes of health inequities are complex, arising from the interaction of various political, economic, and social factors. Still, CSDH was optimistic that the vexing issues surrounding health inequity could be resolved as long as its call for urgent action is heeded (CSDH, 2008). CSDH outlined the various actions, approaches, and recommendations that aspire to close the health gap in a generation. As one goes through the voluminous report, one quickly realizes that no single solution can address today's global health inequity. However, one thing is clear. The health inequities are human-made (Friel, 2017).

This is encouraging. Since the problems are human-made, potential solutions exist if there is the intent and resolve to find them. Therefore, to appreciate how one can help enhance global health's sustainability, one must understand the various components that make up the healthcare systems, the governance organizations, the statistics, and the economic operating models that help evaluate progress. It is equally important to conduct a stakeholder analysis and identify the gaps that hinder the operating models' efficiencies.

2.2.1 Major Global Health Organizations

Global health governance (GHG) refers to the use of formal and informal institutions, rules, and processes by states, intergovernmental organizations, and non-state actors to deal with global health challenges effectively (Fidler, 2010). GHG has four main functions: (1) the production of global goods, guidelines, polices, research, and technologies, (2) the management of external threats, (3) the facilitation of global solidarity, and (4) a stewardship function. Health equity is "the absence of unfair, avoidable, or remediable differences among groups of people, whether those groups are defined socially, economically, demographically, geographically, or by other stratifiers (e.g., sex, gender, ethnicity, disability, or sexual orientation)". GHG has a moral and functional role in achieving health equity so that every human is entitled to health. Considering the above GHG functions, health equity could be addressed through each and every one of them, as health equity is an essential aspect of responsible governance (Kheir-Mataria et al., 2022). As per the information compiled by the International Medical Volunteers Association (IMVA), the various global organizations that provide long-term healthcare can be divided into a few broad categories: multilateral organizations (or inter-governmental organizations—IGOs), bilateral organizations, public-private partnerships (PPPs), and non-state entities (IMVA, 2022).

Multilateral organizations receive funding from multiple governments (and non-governmental sources) and distribute it to many countries. Key multilateral organizations include agencies such as WHO, the United Nations Children's Fund (UNICEF), the World Food Program (WFP), the Coalition for Epidemic Preparedness Innovations (CEPI), and the UN Population Fund (UNFPA). In particular, WHO is the directing and coordinating authority for health within the UN system. It is responsible for providing leadership on global health matters, shaping the health research agenda, setting norms and standards, articulating evidence-based policy options, providing technical support to countries, and monitoring and assessing health trends. Working through offices in more than 150 countries, WHO staff work side-by-side with governments and other partners to ensure the highest attainable level of health for all people.

Bilateral agencies are governmental agencies in a single country that provide aid to developing countries. This aid is more specifically targeted than multilateral aid. Most prominent of these is the US Agency for International Development (USAID), which was created in 1961.[2] It brought together several existing foreign assistance organizations and programs. Until then, a single US agency had never been charged with foreign economic development. Most industrialized nations have a similar governmental agency. For example, the Japan Bank for International Cooperation (JBIC), the United Kingdom's Foreign, Commonwealth & Development Office (FCDO), and the Danish International Development Agency (DANIDA) all

[2] In 1961, President John F. Kennedy signed the Foreign Assistance Act into law and created USAID by executive order.

undertake development aid programs, many of which include health-related initiatives. These agencies may give aid directly to the recipient country's government or channel it via contracts with non-governmental organizations (NGOs) that can work effectively in partnership with local communities (PBS, 2024).

A public-private partnership (PPP) is a collaborative organizational structure where public, private, and non-profit partners agree to share risks, resources, and decision-making authority. Healthcare PPPs can be particularly useful in low- and middle-income countries, where large percentages of the public cannot afford to pay a market rate for healthcare services (Trager & Simon, 2020). Examples of PPPs include the Global Fund to Fight AIDS, Tuberculosis, and Malaria (commonly known as the Global Fund) and the Vaccine Alliance (Gavi). The Global Fund was created in 2002 to raise, manage, and invest money to respond to three of the deadliest infectious diseases that the world has ever known. Gavi was founded in 2000 as a global health PPP whose mission is to save children's lives and protect people's health by increasing access to immunization in the poorest countries where preventable but deadly diseases are most prevalent. An overview of Gavi is shown in Fig. 2.3.

Non-state entities include organizations and individuals not affiliated with, directed by, or funded through the government. These include corporations, philanthropic foundations, private financial institutions, and NGOs. Together, these non-state entities provide significant external health aid to developing countries. For example, The Bill and Melinda Gates Foundation,[3] formally launched in 2000, has been a critical entrant into the global health sphere and has often served as a catalytic force for new financing, innovation, and implementation approaches. The Gates Foundation spends more each year on global health than any government, except for the US and Britain; it accounted for almost 8% of all development assistance for health in 2016 (Michaud, 2018). Similarly, numerous NGOs operating throughout the world provide critical healthcare support. One of the most influential NGOs in long-term international health is Oxfam,[4] a global organization that fights inequality to end poverty and injustice.

2.2.2 *Financing of Global Health*

With so many organizations working toward solutions, one may ask why global healthcare is still a concern? The world population has indeed increased but so has our knowledge of optimizing our resources and our understanding of diseases. We have made significant progress in solving various issues, such as hygiene and sanitation, and have provided excellent medications that help improve the lives of

[3]The Bill & Melinda Gates Foundation is an American private foundation founded by Bill Gates and Melinda French Gates. Based in Seattle, Washington, it is reported to be the second largest charitable foundation in the world, holding $69 billion in assets as of 2020.

[4]The name "Oxfam" comes from the Oxford Committee for Famine Relief, founded in Britain in 1942.

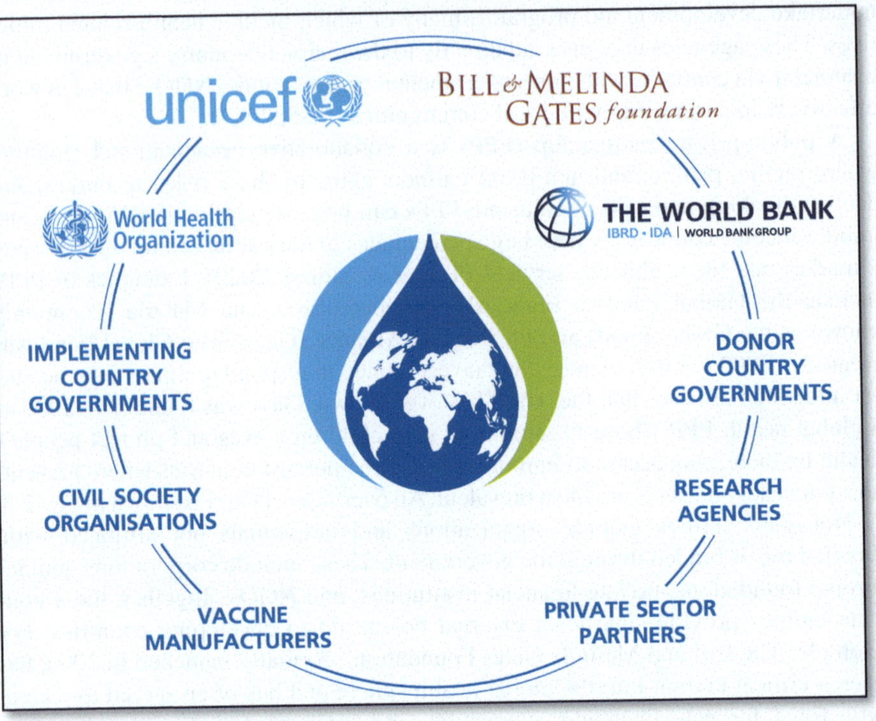

Fig. 2.3 Overview of Gavi: The vaccine alliance. (Source: Gavi) (https://www.gavi.org)

patients suffering from various diseases. Then where is the problem? One way to evaluate this is by examining how much resources are dedicated to financing global health.

The Institute for Health Metrics and Evaluation (IHME) is an independent research organization at the University of Washington that provides rigorous and comparable measurements of the world's most important health problems and evaluates the strategies used to address them. IHME makes this information widely available, so policymakers have the evidence they need to make informed decisions about allocating resources to improve population health. In its annual report, *Financing Global Health*, IHME covers various health focus areas, one of them being the tracking of global health spending. In its 2020 report, IHME tracks spending on global health but specifically highlights development assistance for health support during a pandemic, and that fact—that COVID-19 continues to challenge the world—underscores the importance of tracking resources available for health. However, in its report, IHME also provides an overview of global health spending before COVID-19.

An evaluation of this data is critical as it highlights the discrepancy in health spending worldwide when income is considered. As per the report, in 2018, total health spending reached $8.5 trillion. With $3.6 trillion in 2018 total health

Table 2.1 Total health spending in 2018 (IHME, 2021)

World bank's income group category	Health spending per person (2020 USD)	Health spending per GDP[a]	Government health spending per total health spending
High-income	$5553	12.4%	61.5%
Upper-middle-income	$478	5.6%	55.6%
Lower-middle-income	$87	4.1%	33.0%
Low income	$35	5.0%	24.4%

[a]Gross Domestic Product (GDP) represents the total dollar value of all goods and services produced over a specific period, often referred to as the size of the economy

spending, the United States was the leading global spender, with well over a third of all resources spent on health. Table 2.1 shows the spending split by income group. As can be seen, most health spending occurred in high-income countries, with health spending per person reaching $5553 in 2018, far more than spending per person in low-income countries (IHME, 2021). Other essential patterns illustrated in Table 2.1 include that the percentage of government health spending is significantly less in the low-income countries, thereby leading to the highest dependency on Development Assistance for Health (DAH).[5]

Another important dataset from IHME shows how the total global health spending grew between 1995 and 2018 (Fig. 2.4), with spending disaggregated into four sources: DAH, prepaid private health spending, out-of-pocket spending, and government health spending. Prepaid private spending covers health spending on private insurance premiums and through domestic non-governmental organizations. Out-of-pocket spending includes all health spending not paid in advance. Finally, government health spending is defined as spending on health from all levels of government, in public and private facilities.

Tracking each source of spending is essential for many reasons. Understanding how government spending is allocated helps to reveal what funding is needed to build robust public health systems, while prepaid spending (government or private) is essential for reducing the possibility of catastrophic health spending on the part of individuals. Robust prepaid spending is also essential for reaching universal health coverage. Understanding out-of-pocket spending, meanwhile, tells us the opposite: how much health spending falls on individuals. Last but certainly not least, tracking DAH shows which external sources of funding support which health focus areas and how that money is disbursed. Just as importantly, tracking DAH can show the degree to which lower-income countries rely on external funding, which is

[5]DAH tracks the financial and in-kind resources that are transferred through international development agencies (such as UNICEF, the United Kingdom's Department for International Development, or the Bill & Melinda Gates Foundation) to low- and middle-income countries with the primary purpose of maintaining or improving health.

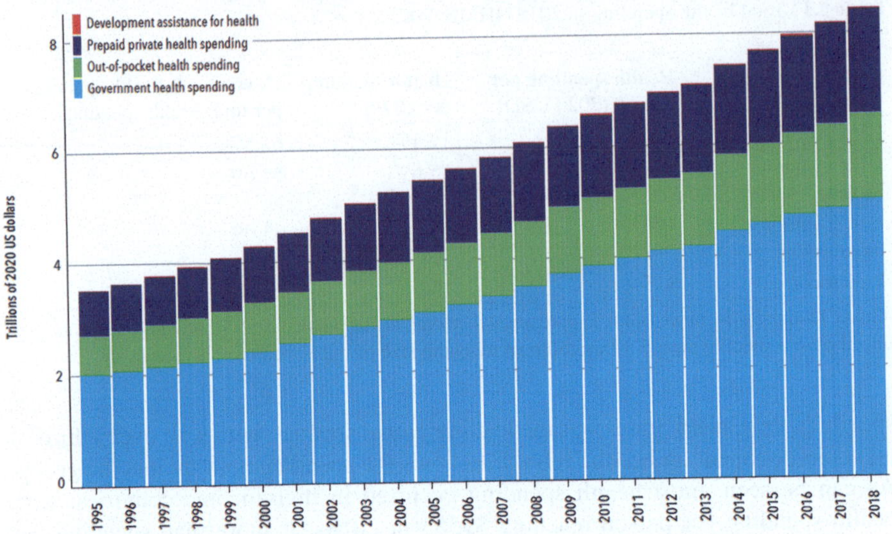

Fig. 2.4 Total health spending by source of financing (1995–2018). (IHME, 2021)

crucial to understand when planning for future development and, eventually, sustainability (IHME, 2021).

Upon carefully reviewing Fig. 2.4, it is clear that since 1995, the relative fraction of each financing source has remained relatively constant. For example, prepaid private health spending went from 22.7% of total spending in 1995 to 22.2% in 2018; out-of-pocket spending went from 19.5% in 1995 to 18.1% in 2018. The DAH fraction is significantly small, with 0.31% of total spending on health in 1995 and 0.46% in 2018. As expected, government health spending remains the leading source of spending, consistently being greater than 55% of total spending (IHME, 2021). However, as expected, this government health spending is unequal among all the countries. In the same report, IHME also evaluated government health spending as a fraction of the total health spending. This vital evaluation provides insights into which countries are on the path to financial risk protection and, therefore, are less dependent on donor funding and out-of-pocket spending. Therefore, government health spending is also a surrogate measurement of the robustness of a country's health systems. As shown in Fig. 2.5, the gap between government health spending in the various countries cannot be wider! In general, all countries have committed substantial resources toward healthcare (as discussed in Table 1.2). However, as seen in Fig. 2.5, a higher GDP does not efficiently translate into higher government health spending. This is due to various socioeconomic, political, and resource management factors. It should be no surprise that health inequality continues to remain a frustrating global problem.

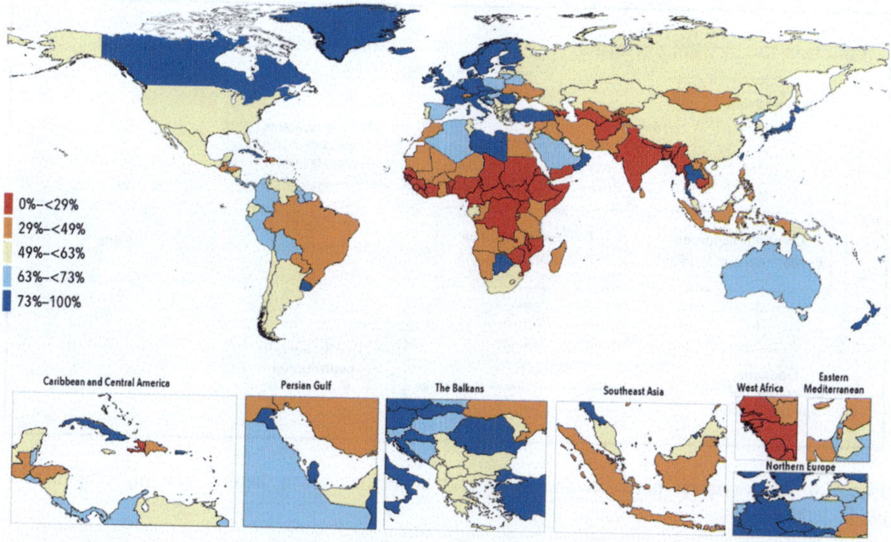

Fig. 2.5 Government spending as a fraction of total health spending in 2018. (IHME, 2021)

2.3 Collated Efforts to Address Global Health Inequality

WHO estimates that half of the world's population cannot access the medical services they need. Over 930 million people spend at least 10% of their household income on medical services, and 100 million people are driven into poverty each year through out-of-pocket health spending (WHO, 2022). Even in developed countries like the United States, medical debt is the principal cause of personal bankruptcy. In 2008, researchers concluded that illness or medical bills had contributed to 62% of bankruptcies the previous year. Many of these people had some form of health insurance, but the benefits offered were insufficient to protect them against high out-of-pocket expenses (WHO, 2010). Also, questions about the financial viability of publicly funded healthcare systems have arisen due to the challenging financial climate in many countries. Given its importance, one can imagine how addressing health inequality is a significant concern of all local, state, national, and global agencies. Consequently, there is increasing interest among researchers, evaluators, funders, and community partners to define, understand, and provide a framework for the sustainability of effective public health systems (Garzón-Orjuela et al., 2020; Walugembe et al., 2019). Based on the literature, a few attributes can help in the attainabililty of a successful and sustainable health system.

Indisputably, a fundamental goal of any health system is the improvement of the health of its population. Along the same lines, a viable health system framework links health system functions (namely factors that influence the health system's performance) to health system objectives, as shown in Fig. 2.6 (Papanicolas et al., 2022). A successful health system has three general attributes: (1) a ***healthy***

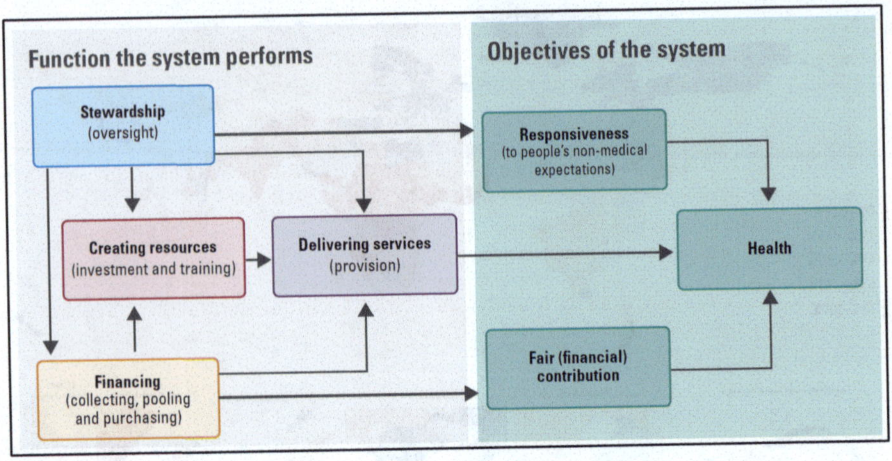

Fig. 2.6 Relationship between functions and objectives of a health system. (Papanicolas et al., 2022)

population that attains the highest level of health possible; (2) ***superior care*** that is effective, safe, timely, patient-centered, equitable, and efficient; and (3) ***fairness of the system*** towards patients, health professionals, institutions, and businesses supporting and delivering care. Similarly, a sustainable health system also has three key attributes: (1) ***affordability*** for patients, employers, and the government; (2) ***acceptability*** to key constituents, including patients and health professionals; and (3) ***adaptability*** to changing healthcare needs (Fineberg, 2012). The development of successful and sustainable health systems is a desirable goal for all countries. However, given its mandate to provide leadership on matters critical to health and engage in partnerships where joint action is needed, the UN and the WHO play a vital role in monitoring the global health situation and assessing health trends. This helps shape the research agenda and stimulate the data generation, translation, and dissemination of valuable knowledge to all member states, thereby providing technical support, catalyzing change, and building institutional capacity.

2.3.1 Sustainable Development Goals (SDG)

In 2000, the then UN Secretary-General Kofi Annan introduced eight Millennium Development Goals (MDGs) that were designed to improve the lives of the world's poor. The MDGs were adopted by 189 nations and had notable successes. Despite the absence of any legally binding framework, the MDGs generated considerable public and policy support nationally and among international agencies and foundations, ensuring the efficient channeling of significant funds (Griggs et al., 2014). In addition, the MDGs played a historic and transformative role in streamlining global attention, resource, and programmatic action onto three health and development

Fig. 2.7 Sustainable development goals of United Nations. (Source: UN)

priorities: the reduction of child mortality (MDG 4), the improvement of maternal health (MDG 5), and combatting HIV/AIDS, malaria and other diseases (MDG 6) (Brolan, 2017). In 2016, the 17 Sustainable Development Goals (SDGs) of the 2030 Agenda for Sustainable Development—adopted by world leaders in September 2015 at a historic UN Summit—officially came into force. The SDGs, also known as Global Goals, build on the success of the MDGs and aim to go further to end all forms of poverty (Fig. 2.7).

With these new Goals that universally apply to all, countries will mobilize efforts to end all forms of poverty, fight inequalities, and tackle climate change while ensuring that no one is left behind. The new Goals are unique in that they call for action by all countries, poor, rich, and middle-income, to promote prosperity while protecting the planet. They recognize that ending poverty must go together with strategies that build economic growth and address various social needs, including education, health, social protection, and job opportunities while tackling climate change and environmental protection. While the SDGs are not legally binding, governments must take ownership and establish national frameworks to achieve the 17 Goals. Countries are primarily responsible for following up and reviewing the progress made in implementing the goals, which will require quality, accessible, and timely data collection. Regional follow-up and review will be based on national-level analyses and contribute to follow-up and review at the global level.

2.3.2 SDG 3

Global health was represented in the 17 SDGs with a single goal, SDG 3: *Ensure healthy lives and promote well-being for all at all ages*. SDG 3 has nine targets (Table 2.2) and four means of implementation that significantly assist in achieving sustainable global health. All of these targets are admirable and can make a substantial difference for humanity. However, the reality is different from ideality. In its latest report, the UN acknowledges that there has been some progress in improving global health in recent years. For example, 146 out of 200 countries or areas have already met or are on track to meet the SDG target on under-5 mortality. Effective HIV treatment has cut global AIDS-related deaths by 52% since 2010, and at least one neglected tropical disease has been eliminated in 47 countries (as discussed in Sect. 1.2.2).

However, insufficient progress has been made in other areas, such as reducing maternal mortality and expanding universal health coverage. Globally, approximately 800 women died every day from pregnancy or childbirth in 2020. Furthermore, 381 million people were pushed or further pushed into extreme poverty in 2019 due to out-of-pocket payments for health. The COVID-19 pandemic and ongoing crises have impeded progress towards Goal 3. Childhood vaccinations have experienced the largest decline in three decades, and tuberculosis and malaria deaths have increased compared with pre-pandemic levels. To overcome these

Table 2.2 Targets of UN's SDG #3 (Good health and well-being)

Target	Details
3.1	By 2030, reduce the global maternal mortality ratio to less than 70 per 100,000 live births.
3.2	By 2030, end preventable deaths of newborns and children under 5 years of age, with all countries aiming to reduce neonatal mortality to at least as low as 12 per 1000 live births and under-5 mortality to at least as low as 25 per 1000 live births.
3.3	By 2030, end the epidemics of AIDS, tuberculosis, malaria, and neglected tropical diseases and combat hepatitis, water-borne diseases, and other communicable diseases.
3.4	By 2030, reduce by one-third premature mortality from non-communicable diseases through prevention and treatment and promote mental health and well-being.
3.5	Strengthen prevention and treatment of substance abuse, including narcotic drug abuse and harmful use of alcohol.
3.6	By 2020, half the number of global deaths and injuries from road traffic accidents
3.7	By 2030, ensure universal access to sexual and reproductive healthcare services, including family planning, information and education, and the integration of reproductive health into national strategies and programs
3.8	Achieve universal health coverage (UHC), including financial risk protection, access to quality essential healthcare services, and access to safe, effective, quality, and affordable essential medicines and vaccines for all
3.9	By 2030, substantially reduce the number of deaths and illnesses from hazardous chemicals and air, water, and soil pollution and contamination

Source: UN

setbacks and address long-standing healthcare shortcomings, increased investment in health systems is needed to support countries in their recovery and build resilience against future health threats (UN, 2023).

2.3.3 *Universal Health Coverage (UHC)*

SDG Target 3.8 (achieving Universal Health Coverage or UHC) has a vast potential to reduce the financial hardship faced by underserved people and communities. Including UHC in the SDGs presents an opportunity to promote a comprehensive and coherent approach to health and strengthen healthcare systems. Countries that advance UHC in their policy-setting agendas will make progress toward other health-related targets. UHC has, therefore, become a primary goal for health reform in many countries and a priority objective of WHO. UHC means that all people have access to the medical services they need, when and where they need them, without financial hardship or catastrophic health spending. It includes a full range of essential services, from health promotion to prevention, treatment, rehabilitation, and palliative care. UHC does not mean free access to every possible health service for every person. Every country has a different path to achieving UHC and decides what to cover based on the needs of its people and the resources at hand. However, it emphasizes the importance of access to health services and information as a fundamental human right (WHO, 2022).

WHO identifies achieving universal coverage as a strategic priority, with the goal of 1 billion more people benefiting from UHC by 2023. In countries with fragile health systems, WHO focuses on technical assistance to build national institutions and service delivery to fill critical gaps in emergencies. In more robust health system settings, WHO drives public health impact towards health coverage for all through policy dialogue for the systems of the future and strategic support to improve performance. This work is supported by normative guidance and agreements, data, research, innovation, and leadership in diplomacy, advocacy, gender equality, health equity and human rights, multisectoral action, and finance.

Progress towards reaching SDG Target 3.8 (as listed in Table 2.2) on UHC is tracked using two separate metrics: (1) SDG indicator 3.8.1 captures the coverage of essential health services irrespective of people's living standards, and (2) SDG indicator 3.8.2 captures the financial protection dimension of UHC such that use of health services should not lead to financial hardship or catastrophic health spending. Both indicators must be measured together to capture the complete picture. WHO is the designated custodian agency for both SDG 3.8 indicators, with the United Nations Children's Fund (UNICEF), United Nations Population Fund (UNFPA), and United Nations Department of Economic and Social Affairs (UN DESA) Population Division as partner agencies for 3.8.1, and with the World Bank for 3.8.2 (WHO, 2021).

2.3.4 *Coverage of Essential Health Services (SDG Indicator 3.8.1)*

SDG indicator 3.8.1 (coverage of essential health services) is defined as the average coverage of essential services based on tracer interventions that include reproductive, maternal, newborn, and child health, infectious diseases, non-communicable diseases, and service capacity and access, among the general and the most disadvantaged population. The UHC Service Coverage Index (SCI[6]) is reported on a unitless scale of 0 to 100, computed as the geometric mean of 14 health service coverage tracer indicators. These indicators are meant to indicate service coverage and should not be interpreted as a complete or exhaustive list of the health services or interventions required to achieve UHC. Based on WHO's data from 2019, the global distribution of UHC SCI is shown in Fig. 2.8 (WHO, 2022).

WHO's 2019 Global Monitoring Report on UHC offers good and bad news. On the one hand, the world continues to progress in healthcare access, particularly in population coverage with maternal and child health services and services preventing and responding to infectious diseases. On the other hand, progress is too slow to reach the SDG coverage target and will need to accelerate, particularly in Africa and low-income countries. Progress must be markedly accelerated to reach UHC (SDG target 3.8.1). In 2017, nearly half the world's people (33% to 49%) were covered by essential health services. The number to be covered in the SDG era (2015–2030) is projected to increase by 1.1 to 2.0 billion, but the trend is offset by population growth, so the percentage of people covered rises more slowly. If current trends continue to 2030, only 39% to 63% of the global population will be covered by essential health services. Coverage needs to at least double to reach the SDG target of UHC by 2030 (WHO, 2021).

2.3.5 *Catastrophic Health Spending (SDG Indicator 3.8.2)*

Financial protection is a crucial health system objective and an essential dimension of UHC. Financial protection means everyone can obtain the health services they need without experiencing financial hardship. SDG indicator 3.8.2 defines catastrophic health spending as out-of-pocket health spending exceeding 10% and 25% of the household budget (total consumption or income). Out-of-pocket health spending (out-of-pocket payments) is defined as household spending on medicines, health products, outpatient and inpatient care services (including dental care), and other health services (such as medical laboratory services) that are not reimbursed

[6] UHC SCI, which is the official measure for SDG indicator 3.8.1, was developed as part of a multiyear process that included global reviews, country case studies, consultations with health officials, and a formal WHO country consultation with member states in 2017.

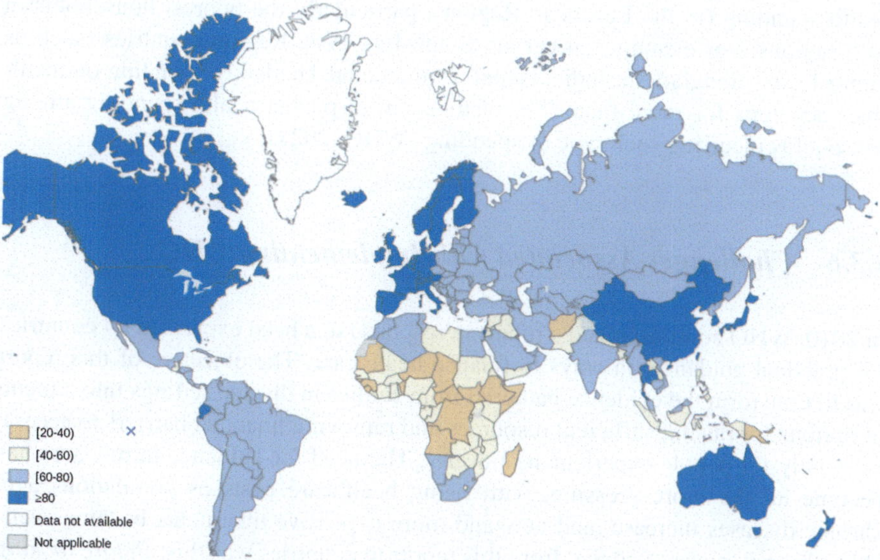

Fig. 2.8 UHC service coverage index for 2019 (WHO, 2022)

by a third party (such as the government, a health insurance fund or a private insurance company). It excludes household spending on health insurance premiums (WHO, 2021).

As per WHO's 2019 Global Monitoring Report, in 2015, about 930 million people incurred out-of-pocket health spending exceeding 10% of the household budget; for about 210 million people, out-of-pocket health spending exceeded 25% of the household budget. Globally, catastrophic health spending increased continuously between 2000 and 2015, as tracked by SDG indicator 3.8.2. The share of the world's population with out-of-pocket health spending exceeding 10% of the household budget went from 9.4% in 2000 to 12.7% in 2015, and the share with out-of-pocket health spending exceeding 25% of the household budget from 1.7% in 2000 to 2.9% in 2015. Between 2000 and 2015, the largest concentration of the world population with out-of-pocket health spending exceeding 10% and 25% of their household budget shifted from low-income to middle-income countries. Between 2000 and 2015, out-of-pocket health spending increased global poverty at varying paces depending on the poverty line (WHO, 2021).

The same report shared data to discuss which health services drive financial hardship. Evidence from the South-East Asia Region, the European Region, and selected countries, mainly in Africa, suggests that medicine accounts for the largest share of out-of-pocket health spending among people incurring any out-of-pocket health spending (in South-East Asia) and among households with catastrophic health spending (in the European Region), particularly the poorest households in both regions. For example, in certain South-East Asia Region countries (such as Timor-Leste, Bangladesh, India, Nepal, Bhutan, and Thailand), spending on medicines accounts for more than 75% of total out-of-pocket health spending among households incurring such type of spending (WHO, 2021).

2.3.6 Challenges Associated with Implementing UHC

In 2010, WHO commissioned a report in response to a need expressed by countries for practical guidance on ways to finance healthcare. The objective of this report was to transform the evidence gathered from studies in diverse settings into a menu of options for raising sufficient resources and removing financial barriers to access, especially for people experiencing poverty. The need for guidance in this area has become all the more pressing, with rising healthcare costs as populations age, chronic diseases increase, and new and more expensive treatments become available. One key piece of advice from this report to countries was this: *Before looking for places to cut spending on* healthcare, *look first for opportunities to improve efficiency*. All health systems, everywhere, could use resources better, whether through better procurement practices, broader use of generic products, better incentives for providers, or streamlined financing and administrative procedures. The report estimates that from 20% to 40% of all health spending is currently wasted through inefficiency and points to 10 specific areas where better policies and practices could increase the impact of expenditures, sometimes dramatically. Investing these resources more wisely can help countries move closer to universal coverage without increasing spending (WHO, 2010).

Progress towards UHC depends on raising adequate funds from a sufficiently large pool of individuals, supplemented where necessary with donor support and general government revenues, and spending these funds on the services a population needs. In general, the bigger the pool, the better able it is to cope with the financial risks. As per the WHO 2010 report, there is no single way to develop a financing system to achieve universal coverage. All countries must make choices and trade-offs, particularly in how pooled funds are used. It is a constant challenge to balance priorities: funds often remain scarce, yet people demand more, and the technologies for improving health are constantly expanding. Such conflicts force policy-makers to make trade-offs in three core areas (Fig. 2.9): the proportion of the population to be covered, the range of services to be made available, and the proportion of the total costs to be met. The box labeled "current pooled funds" in Fig. 2.9 depicts a hypothetical situation where, in order to get closer to UHC, the country would need to extend coverage to more people, offer more services, and pay a greater part of the costs from pooled funds (WHO, 2010). As can be imagined, a few ways by which a higher rate of UHC can be achieved is by either allocating more money for health or by getting more health for the same amount of money by using resources wisely.

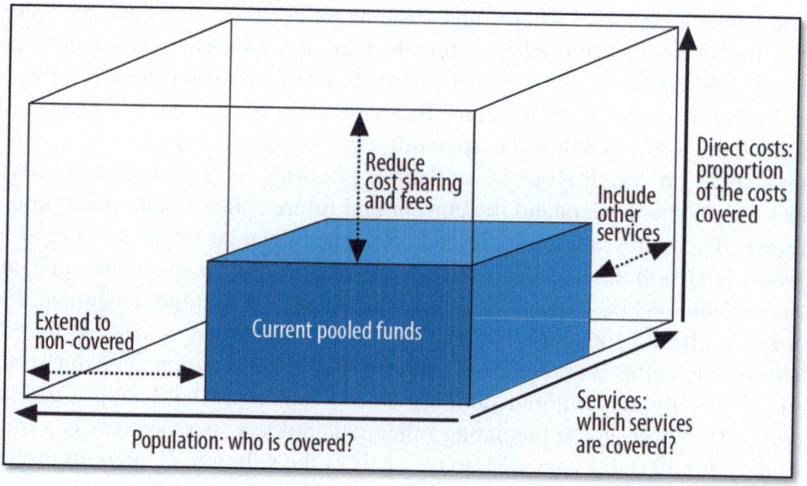

Fig. 2.9 Three dimensions of UHC. (WHO, 2010)

The first approach of allocating more money for health is not always guaranteed. Countries must continually adapt their financing systems to raise sufficient funds for their health systems. Many high-income countries are facing a decline in the proportion of their working-age population and having to consider alternatives to traditional sources of revenue in the form of income taxes and health insurance contributions from workers and their employers. In many lower-income countries, more people work in the informal than the formal sector, making collecting income taxes and wage-based health insurance contributions challenging. The second approach of using healthcare resources wisely is more promising for extending the UHC. All countries can do something, many of them a great deal, to improve the efficiency of their health systems, thereby releasing resources that could be used to cover more people, more services, and more of the costs. There are many opportunities for efficiency gains. This does not mean simply cutting costs. Efficiency measures the quality and quantity of output (i.e., health outcomes or services) for a given input level (i.e., cost). So, efficiency gains could help to contain costs—an essential objective in many countries—by reducing service delivery costs. However, no one wants to contain costs by reducing health outcomes, so seeking efficiency gains should also be seen as a means of extending coverage for the exact cost (*please see* Sect. 8.3 *for further discussion on this topic*).

2.4 Summary—Be a Part of the Solution

Let us face it. There is no magical solution to bridging the health inequity gap. After all, the health inequity gap is a function of numerous socioeconomic, historical, and cultural factors that shape our society. However, not trying to be a part of the solution is not acceptable! Robust and sustainable health systems are necessary to build resilience against future health threats. One way to strengthen the health systems and offer financial protection to patients is by reducing household out-of-pocket health spending. Therefore, an effective strategy to ensure UHC advancement goals are met is by tackling the spending on medicines. Information about the amounts spent on which types of medicines provides an opportunity to compare countries and better understand the current and future state of attempts to improve healthcare. This is a complex topic and requires a detailed understanding of how pharmaceutical policies are designed to improve access to medicines and make them affordable. While it may not be possible to eliminate impoverishment due to out-of-pocket health spending using a relative poverty line, it is possible to reduce it. Such a noble, although polarizing, exercise requires us to appreciate the perspective of all stakeholders without jumping to conclusions or hardening a defensive posture(s). However, not appreciating other stakeholders' perspectives is a missed opportunity for all those who wish to be a part of the solution. ***Is such an inclusive dialogue not a win-win scenario for society?***

References

Birch, S., Murphy, G. T., MacKenzie, A., & Cumming, J. (2015). In place of fear: Aligning health care planning with system objectives to achieve financial sustainability. *Journal of Health Services Research & Policy, 20*(2), 109–114. http://www.jstor.org/stable/26751551

Brodit, K., & Radwan, H. (2018). Global health equity. *Journal of Best Practices in Health Professions Diversity, 11*(2), 150–154. https://www.jstor.org/stable/26894215

Brolan, C. E. (2017). The right to health and the post-2015 health and sustainable development goal agenda. In *Healthcare as a human rights issue* (pp. 293–320). Transcript Verlag. http://www.jstor.org/stable/j.ctv1fxf7w.13

Fidler, D. P. (2010). *The challenges of global health governance*. http://www.jstor.org/stable/resrep24171

Fineberg, H. V. (2012). A successful and sustainable health system—How to get there from here. *New England Journal of Medicine, 366*(11), 1020–1027. https://doi.org/10.1056/NEJMsa1114777

Friel, S. (2017). Governance, regulation and health equity. In *Regulatory theory* (pp. 573–590). ANU Press. http://www.jstor.org/stable/j.ctt1q1crtm.45

Friendly, M. (2009). The golden age of statistical graphics. *Statistical Science, 23*. https://doi.org/10.1214/08-STS268

Garzón-Orjuela, N., Samacá-Samacá, D. F., Luque Angulo, S. C., Mendes Abdala, C. V., Reveiz, L., & Eslava-Schmalbach, J. (2020). An overview of reviews on strategies to reduce health inequalities. *International Journal for Equity in Health, 19*(1), 192. https://doi.org/10.1186/s12939-020-01299-w

Griggs, D., Smith, M. S., Rockström, J., Öhman, M. C., Gaffney, O., Glaser, G., Kanie, N., Noble, I., Steffen, W., & Shyamsundar, P. (2014). An integrated framework for sustainable development goals. *Ecology and Society, 19*(4) http://www.jstor.org/stable/26269703

Hancock, T. (1993). Health, human development and the community ecosystem: Three ecological models. *Health Promotion International, 8*(1), 41–47. http://www.jstor.org/stable/45152125

IHME. (2021). *Financing global health 2020: The impact of COVID-19*. https://www.healthdata.org/sites/default/files/files/policy_report/FGH/2021/FGH_2020_full-report.pdf

IMVA. (2022). *The major international health organisations*. Retrieved June 3, 2022 from http://www.imva.org/Pages/orgfrm.htm
Jaba, E., Balan, C. B., & Robu, I.-B. (2014). The relationship between life expectancy at birth and health expenditures estimated by a cross-country and time-series analysis. *Procedia Economics and Finance, 15*(14), 108–114.
Jeggo, M., Arabena, K., & Mackenzie, J. S. (2019). One health and global security into the future. In M. Walton (Ed.), *One planet, one health* (pp. 21–52). Sydney University Press. https://doi.org/10.2307/j.ctvggx2kn.7
Kheir-Mataria, W. A. E., El-Fawal, H., Bhuiyan, S., & Chun, S. (2022). Global health governance and health equity in the context of COVID-19: A scoping review. *Healthcare (Basel), 10*(3). https://doi.org/10.3390/healthcare10030540
Kjærgård, B., Land, B., & Bransholm Pedersen, K. (2013). Health and sustainability. *Health Promotion International, 29*(3), 558–568. https://doi.org/10.1093/heapro/das071
Magnello, E. (2010). Florence nightingale: A Victorian statistician. *Mathematics in School, 39*(3), 20–21. http://www.jstor.org/stable/20799090
Michaud, J. (2018). Global health progress and challenges. *Great Decisions*, 91–102. https://www.jstor.org/stable/26593700
Papanicolas, I., Rajan, D., Karanikolos, M., Soucat, A., & Figueras, J. (2022). *Health system performance assessment: A framework for policy analysis*. World Health Organization.
PBS. (2024). *Politics and global health*. PBS. Retrieved January 16, 2024 from https://www.pbs.org/wgbh/rxforsurvival/series/politics/who_bilateral.html
Trager, A. M., & Simon, E. (2020). *PPP government guide: Designing healthcare solutions with PPPs*. PPP Initiative Ltd. Retrieved from https://www.hhs.gov/sites/default/files/ppp-government-guide-healthcare.pdf
UN. (2023). *The sustainable development goals report—Special edition*. https://unstats.un.org/sdgs/report/2023/The-Sustainable-Development-Goals-Report-2023.pdf
Walugembe, D. R., Sibbald, S., Le Ber, M. J., & Kothari, A. (2019). Sustainability of public health interventions: Where are the gaps? *Health Research Policy and Systems, 17*(1), 8. https://doi.org/10.1186/s12961-018-0405-y
WHO. (2010). *The world health report: Health systems financing—The path to universal coverage*. Retrieved from https://www.who.int/publications/i/item/9789241564021
WHO. (2021). *Primary health care on the road to universal health coverage: 2019 global monitoring report*. World Health Organization. https://apps.who.int/iris/handle/10665/344057
WHO. (2022). *Universal health care*. WHO. Retrieved August 6, 2022 from https://www.who.int/health-topics/universal-health-coverage#tab=tab_1

Chapter 3
Perspective of Stakeholders: Understanding Viewpoints

Where you stand depends on where you sit

Rufus Miles

Contents

B. Mittal, *Sustainable Global Health Systems and Pharmaceutical Development*, AAPS Advances in the Pharmaceutical Sciences Series 60,
https://doi.org/10.1007/978-3-031-50415-0_3

Abstract The accessibility and affordability of medicines are a primary societal concern. On the same note, pharmaceutical drug pricing is a polarizing issue. Numerous stakeholders have diverse opinions on what pharmaceutical pricing should be like. Similarly, ideas abound in regards to improving access to medications. However, from the pharmaceutical industry's perspective, developing novel drugs is a highly costly endeavor. Fortunately, an opportunity exists for a win-win scenario to be seized upon by proactively seeking out the perspectives of all stakeholders. After all, drug development is a complex team sport, and most of the time, a decision on any aspect of product development cannot be made without understanding its impact on other disciplines' work. From this perspective, one cannot help but appreciate the various sciences working together to achieve a common goal: developing a safe and effective drug product. However, like all things in life, innovations in the field of pharmaceuticals cannot be evaluated in isolation. One must also understand the trials and tribulations of other areas where fundamental research is being conducted and solutions are being developed for humanity.

Keywords Stakeholder analysis · Health insurance · Pharmaceutical pricing · Generics · PROM

3.1 Introduction

Today, the science of drug development is an established field that requires a dedicated understanding of numerous disciples and fosters a symbiotic partnership between multiple subject matter experts. Given the experience that we now have in drug development, the steps taken toward establishing a drug's safety, efficacy, and the process for its commercialization have long been standardized. However, the road to pharmaceutical innovation was and is never easy (*please see* Sect. 6.1.2 *for the historical evolution of modern drugs*). It is lamentable that the hallmarks of progress in the groundbreaking understanding of disease and the development of life-saving medicines are routinely overshadowed by the commercial aspect of the pharmaceutical business. There are justifiable and understandable reasons for such an aversive response from stakeholders. It seems that not only does the public have a poor perception of the industry, but the industry, too, cannot help overcome the narrative when it is accused of chicanery.

Some prominent examples are when, in 2015, Turing Pharmaceuticals raised the price of Daraprim (pyrimethamine) 75 mg dose tablet, a then-62-year-old drug that is used to treat toxoplasmosis[1] and malaria, from \$13.50/tablet to \$750/tablet, a

[1] About one-third of the population is infected with *Toxoplasmosis gondii*, the parasite that causes toxoplasmosis. Most people experience no symptoms. However, toxoplasmosis can cause blindness, seizures, impaired cognition, and even life-threatening problems in infected people with compromised immune systems such as AIDS and certain cancer patients, and even babies born to

5500% increase, illustrating how the absence of competition[2] in the sale of low-volume, low-price drugs can lead to price gouging. Another well-known case is that of when Gilead Sciences launched Sovaldi® (sofosbuvir) in December 2013 to treat hepatitis C and charged $84,000 per course of treatment. Because many hepatitis C patients qualify for Medicare, providing Sovaldi was expected to increase Medicare Part D spending by $3–6 billion annually. In July 2014, Congress opened an investigation into the price of Sovaldi that revealed Gilead's pricing strategy was intended to capture the largest market share at the highest price for as long as possible. In contrast, the price of a course of Sovaldi in Egypt is $900, a price that would make the drug accessible to many more of the 3 million hepatitis C patients in the United States (Halpenny, 2016).

The virtues of patience and maturity to appreciate and empathize with someone else's perspective without dissolving a worthy discussion into a rancorous and unproductive debate are in short supply in recent times. Everyone seems to have strong opinions on any topic, and it is hard to meet in the middle ground. Unfortunately, drug pricing is one such polarizing issue that frequently leads to a frustrating quagmire. It does not have to be! This discussion has multiple stakeholders: healthcare providers, the pharmaceutical industry (both innovative and generic), the government, insurance companies, and patients. An opportunity exists for a win-win scenario to be seized upon by appreciating the perspectives of all stakeholders and trying to resolve issues through ownership of the problem, participation, and debate among stakeholders, and charting a mutually beneficial path forward. Such a trail-blazing opportunity requires empathy and a keen situational awareness of how high drug prices hurt everyone. No matter how divisive the debate becomes, there is no question that the development of drug products for human consumption is complex and challenging but a worthy undertaking for the betterment and advancement of civilization.

Such wholesome, inclusive, and well-intentioned discourses are not new. Society has faced similar difficulties in the past, and new leaders have risen to the challenge. A combination of sound policy, good intent, and appreciation of one's responsibility have helped our society overcome some of humankind's most complex problems. One such positive example was that of the interim years between the two World Wars when the Great Recession gripped the world. It is worth looking at how one of humanity's most challenging crisis was solved and how it ushered in the modern era, for which we all are eternally grateful.

---

women who become infected during pregnancy. For example, 24–47% of *T. gondii*-seropositive AIDS patients developed encephalitis without prophylactic treatment (Halpenny, 2016).

[2] In 2015, with Turing Pharmaceuticals' acquisition of the US marketing rights for Daraprim tablets, Daraprim became a single-source and specialty pharmacy item.

3.1.1 *Philosophical Digression: Perseverance in Profound Adversity*

The Great Depression was a severe worldwide economic event triggered by the US stock market crash on October 29, 1929. During the Great Depression, worldwide GDP fell by an estimated 15%.[3] Devastating effects were seen in rich and poor countries with falling personal income , prices, tax revenues, profits, and prices. International trade plunged by more than 50%, unemployment in the US rose to 23%, and in some countries, it was as high as 33%. Farming communities and rural areas suffered as crop prices fell by about 60%. Faced with plummeting demand and few job alternatives, areas dependent on primary sector industries suffered the most. The Great Depression was the longest, most profound, and most widespread depression of the 20th century and is regularly used as an example of an intense global economic depression. Even today, a mere mention of it conjures up images of a society marred with vast unemployment lines and food rationing. Very few are alive to tell the tale from that time, but historical records provide a sense of the dire situation, its magnitude, and its dilemma. However, adversity is often a catalyst for change.

The suffering of the Great Depression gave birth to new thinking. The American economist and historian Dr. Edwin F. Gay[4] wrote in 1932: "*The economic losses of the World War and the present depression, in their full volume and extent, are incalculable. The gravity of the present situation lies not merely in the widespread suffering, vast as that is, but in the questions, it excites concerning the fundamental strength and character of our economic structure and in the series of decisions that must be made, indeed, are now in the process of making. These decisions are much more difficult than those of war. In war, the chief problem is set and calls for immediate action, unifying its effect. In such a depression as this, the problem is infinitely complex, decisions are beset by doubt, the action always seems too late or has effects contrary to what was expected, and disunion and disruption have spread as each centrifugal force, seeking to strengthen itself, weakens the whole. War stimulates the full expansion of productive energy, but the deep depression hinders every economic process and discourages even the most sanguine business leaders. There are many confusing prescriptions offered from all sides. But no one, however skilled, really knows the character or the specific cure for what some practitioners diagnose as a wasting disease*" (Gay, 1932). It is indeed a profound analysis of the situation faced by society at that time.

The human spirit is not easily deterred. Like a phoenix rising from the ashes, the leaders of that era rose to face the challenges of the Great Depression. One such approach was the "New Deal", a series of programs, public work projects , financial reforms, and regulations enacted by President Franklin D. Roosevelt in the United

[3] By comparison, worldwide GDP fell by less than 1% during the Great Recession of 2008–2009.

[4] Edwin Francis Gay (1867–1946) was an American economist, Professor of Economic History and first Dean of the Harvard Business School.

States between 1933 and 1939. Major federal programs and agencies included the Civilian Conservation Corps (CCC), the Civil Works Administration (CWA), the Farm Security Administration (FSA), the National Industrial Recovery Act of 1933 (NIRA), and the Social Security Administration (SSA). They supported farmers, the unemployed, the youth, and the elderly. The New Deal included new constraints and safeguards on the banking industry and efforts to re-inflate the economy after prices had fallen sharply. This helped revitalize the US economy and shortened the length of the Great Depression. Today, the legacy of the New Deal remains with programs such as Social Security still in place.

3.1.2 Opportunity for Pharmaceutical Scientists: True Change Starts from Within

The parallels presented by Dr. Gay on the economic losses suffered during the First World War and the Great Depression are interesting to note. As he wrote, war calls for immediate action and is unifying in its effect. We can correlate these thoughts to our present time, where, during the COVID-19 pandemic, all countries mobilized their full financial resources to solve the urgent and immediate crisis. It is almost as if the post-pandemic era is like the Great Depression, where the problem is infinitely complex as decisions are beset by doubt, action always seems too late or has effects contrary to what was expected, and disunion and disruption have spread as each centrifugal force, seeking to strengthen itself, weakens the whole. However, similar to the New Deal, efforts taken after the pandemic can be a galvanizing force to stimulate productive energy that helps overcome the most pressing question of the post-pandemic world—the affordability and accessibility of medicines. The pharmaceutical industry has a huge role to play here.

The pharmaceutical industry is constantly under scrutiny for good and bad reasons alike. On the one hand, it is lauded for its groundbreaking discoveries and for bringing cures to unmet medical needs in the marketplace. The public and the government look towards the pharmaceutical industry for rescue from life-threatening infectious diseases, mental illnesses, and other chronic illnesses. This expectation fosters an environment of mutual respect where the pharmaceutical industry appreciates the trust that is placed in it. Unfortunately, on the other hand, it is criticized on issues regarding patents and access to medicines. Some critics attack the entire patent system, arguing that it leads to high drug prices that few can afford. Other criticisms surround the industry's insistence on including the TRIPS agreement (*please see* Sect. 1.3.2) in the World Trade Organization (WTO) system as an ill-advised move that improperly linked intellectual property and international trade (Desai, 2006).

As discussed in Chap. 2, not only is good health fundamental to our existence, but its pursuit is also a lifelong endeavor. There should be no surprise then that there are strong passions regarding the virtues and banes of the pharmaceutical industry

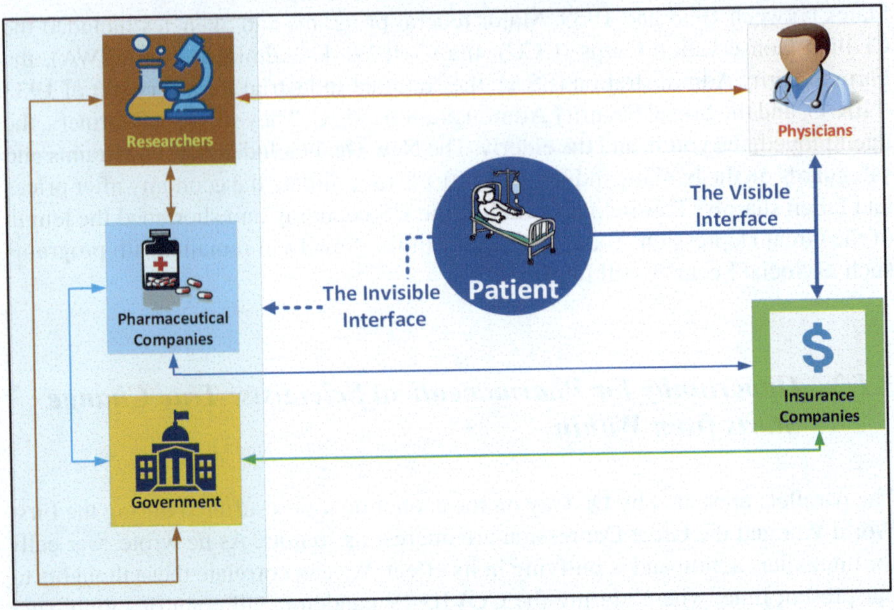

Fig. 3.1 Stakeholders in patients' care

in terms of its role in access to high-quality healthcare. It is, therefore, essential to delve deeper into these discussions by trying to understand the perspectives of all stakeholders in this matter. However, the first and foremost stakeholder is the patient itself. The care of patients is a team effort (Fig. 3.1). This team has two interfaces: a visible one and an invisible one. As anyone who has cared for patients will testify, the visible interface consisting of physicians and insurance companies is the one that (rightly so) gets the most attention in the time of a patient's need. The invisible interface consisting of researchers, pharmaceutical companies, and the government also drives the physicians' options. This invisible interface is the unsung hero of this saga and requires a more profound evaluation.

3.2 Government's Perspective—Improve Access to Medicines

High-quality healthcare is a societal benchmark. It is also a fair expectation that access to good healthcare, irrespective of race, ethnicity, or socioeconomics, is a fundamental human right. Health equity is achieved when it creates fair opportunities for all and eliminates the gaps in health outcomes between different social groups. Better health equity and results can help drive economic growth, bringing peace and prosperity. In that regard, governments have a huge responsibility to provide the best care that they can to their citizens. Numerous researchers have

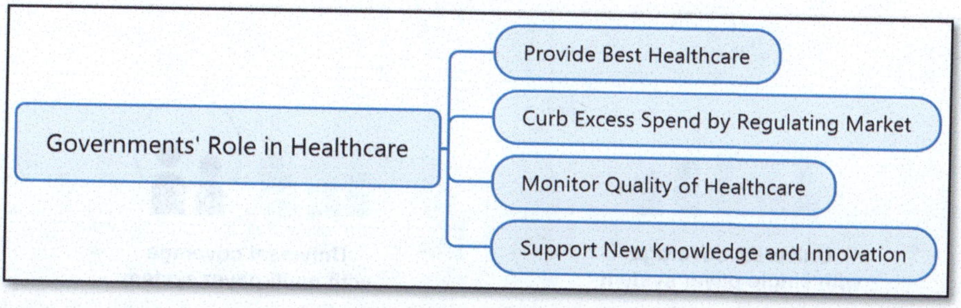

Fig. 3.2 Governments' role in healthcare

discussed the roles of government in improving healthcare quality and safety, and overall, different societies have different expectations of their government's involvement in healthcare (Tang et al., 2004; Kronick, 2016). The core role of government can be broken down into four core areas: (1) provide healthcare, (2) regulate markets, (3) monitor quality, and (4) support innovation (Fig. 3.2). It is, therefore, essential to evaluate the various systems of healthcare governance in the world to set a baseline for better understanding.

3.2.1 *Provide Best Healthcare—A Potpourri of Options*

Healthcare systems around the world differ widely. Countries adopt different systems to provide health care to their citizens, with different levels of government and private sector involvement. The crucial result is that the services exist effectively and safely—and that people can access and afford them. The spectrum of ways in which healthcare is delivered worldwide can be gathered into a few general categories, from universal coverage under fully funded national programs to no coverage at all, requiring individuals to pay for healthcare entirely out of pocket (World101, 2023). A helpful infographic is shown in Fig. 3.3. The numerous existing healthcare systems can also be categorized into five broad groups, as shown in Fig. 3.4. Based on information gathered from multiple sources (Various, 2022), a distribution map of the different healthcare systems and the total number of people they serve can be created, as shown in Fig. 3.5.

- ***Category A (Universal Government-Funded)***: In this system (also known as single-payer healthcare), government-funded healthcare is available to all citizens regardless of income or employment status. Some countries may provide healthcare to non-citizen residents, while some may require them to buy private insurance. A prominent example is the United Kingdom's healthcare system, which covers the whole population via the National Health Service (NHS), which was established in 1948. The NHS is 79% publicly financed from taxes and is operated by the Department of Health. About 20% is paid for by national insurance, and private patients and copayments make up the rest (Thomas, 2021).

Fig. 3.3 Types of healthcare systems. (World101, 2023)

The founding principles were that services should be comprehensive, universal, and accessible at the point of delivery—a health service based on clinical need, not the ability to pay. Each service provides a comprehensive range of health services, free at the point of use for people ordinarily resident in the United Kingdom, apart from dental treatment and optical care.

- ***Category B (Universal Public Insurance)***: Workers have social insurance in countries with Universal Public Insurance. Usually, the government withholds part of their wage, which is divided between employee and employer. People who do not have a legal employment contract and cannot register as unemployed may be ineligible for free healthcare. For example, India has a mixed healthcare system, including public and private service providers. Government workers and most formal employees have their health insurance schemes. However, most private providers are concentrated in urban India, providing secondary and tertiary care services (Chokshi et al., 2016) (Thomas, 2021). The bar chart in Fig. 3.5 shows that nearly 3.4 billion people worldwide are served under this system.

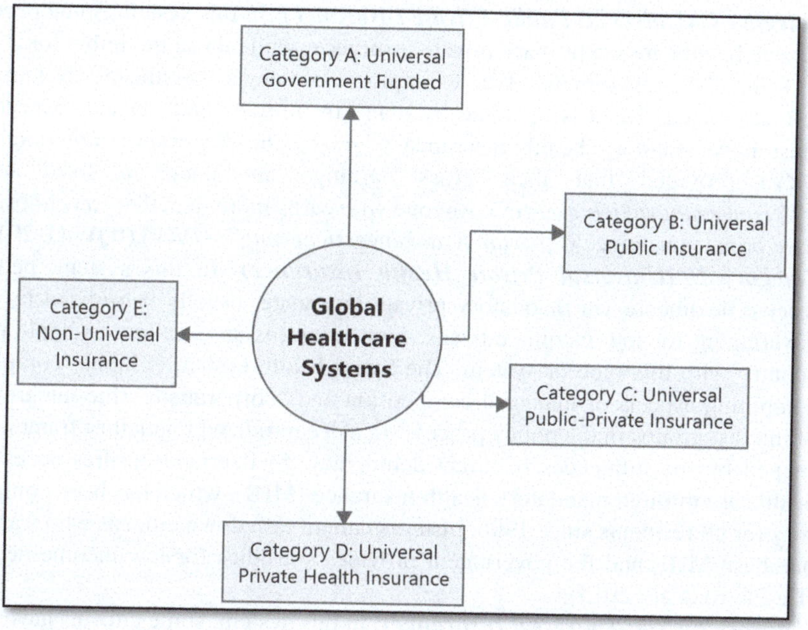

Fig. 3.4 Types of global healthcare systems

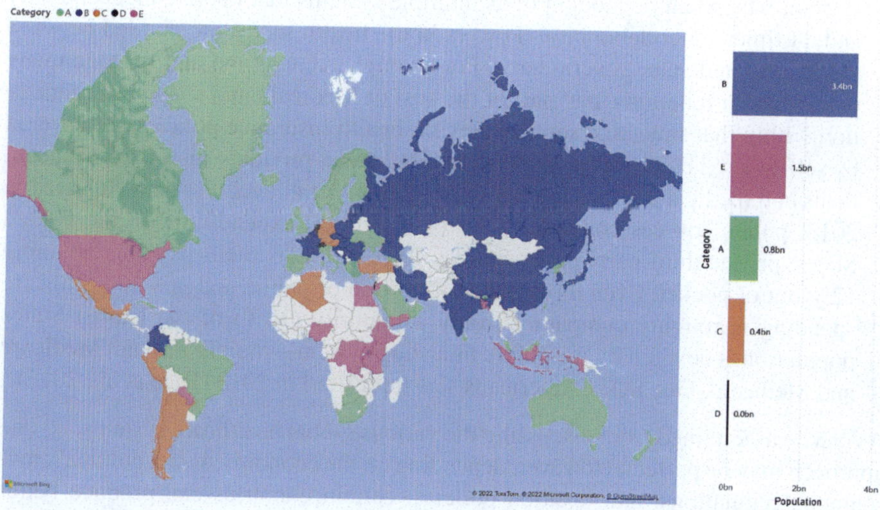

Fig. 3.5 Global healthcare system map and number of people served

- ***Category C (Universal Public-Private Insurance)***: In this system, some people receive healthcare via primary private insurance, while those ineligible for it get it from the government. For example, Germany's healthcare system is self-administered and is operated by many institutions and players. Everyone must have statutory health insurance (*"gesetzliche Krankenversicherung"—GKV*) provided that their gross earnings are under a fixed limit (*"Versicherungspflichtgrenze"*). Anyone who earns more than that can choose to have private insurance (*"private Krankenversicherung"—PKV*) (IQWiG, 2018).
- ***Category D (Universal Private Health Insurance)***: In this system, people receive healthcare via mandatory private insurance, usually subsidized by the government for low-income citizens. Switzerland is an excellent example of a country with this type of system. The Swiss health system is highly complex, combining aspects of managed competition and "corporatism" (the integration of interest groups in the policy process) in a decentralized regulatory framework shaped by the influences of direct democracy. Switzerland ensures access to healthcare through mandatory health insurance (MHI), which has been compulsory for all residents since 1996. Insurers cannot turn down citizens who want to purchase MHI, and the government provides subsidies for low-income people (De Pietro et al., 2015).
- ***Category E (Non-Universal Insurance)***: In this system, some citizens have private health insurance, some are eligible for subsidized public healthcare, and some are not insured at all. A prime example is the United States healthcare system, which can be thought of as multiple systems that operate independently and, at times, in collaboration. Powers in the health sector are divided between the federal and state governments. For example, states fund and manage many public health functions, pay part of the cost of Medicaid and shape its organization within that state, and set the rules for health insurance policies not covered by self-insured employer plans. On the other hand, products such as pharmaceuticals and medical devices are regulated at the federal level. Per data collected in 2013, public sources constitute 48% of healthcare expenditures in the United States, private third-party payer sources 40%, and individuals pay the remaining 12% out of pocket. Even though the proportion of public and private healthcare spending is roughly comparable, only a minority (30%) of the United States population is covered by the public financing system—mainly through Medicare and Medicaid. One in six Americans was uninsured in 2013 (Rice et al., 2013).

As can be ascertained from discussing the various global healthcare systems, no one approach may be perfect. However, depending on the economy, socio-political environment, regulations, and taxation policies, certain countries can produce better healthcare outcomes without spending too many resources. Unfortunately, the United States is not one of those countries, and it can benefit from a healthy and critical review of how its healthcare system is governed.

3.2.2 Address Inefficiencies in Healthcare Systems

Managing healthcare systems is complex and fraught with inefficiency. Most countries, if not all, fail to fully exploit the resources available, whether through poorly executed procurement, irrational medicine use, misallocated and mismanaged human and technical resources, or fragmented financing and administration. WHO postulates that there are ten leading causes of inefficiency, which can advance the cause of UHC when appropriately addressed. These inefficiencies are presented in Table 3.1 (WHO, 2010). There are a few broad categories of inefficiency in the healthcare systems: (1) medicines, (2) healthcare products and services, (3) workers, (4) poor resource allocation, and (5) inefficient implementation of strategies.

It is necessary that before any detailed discussion regarding the pros and cons of the current healthcare market takes place, an evaluation of the inputs and outputs is done. In the United States, the Office of the Actuary (OACT) in CMS catalogs healthcare spending for various categories by the source of funds (for example, PHI, Medicare, Medicaid), by type of service (hospital, physician, prescription drugs, etc.), and by the sponsor (businesses, households, governments). An overview of this data is shown in Fig. 3.6 (CMS, 2021). As can be seen, in 2021, the three major health spending categories were: (1) Hospital Care (31%), (2) Physician and Clinical Services (20%), and (3) Retail Prescription Drugs (9%). In particular, retail prescription drug spending increased by 7.8% to $378.0 billion in 2021, a faster rate than in 2020, when spending increased by 3.7%. The acceleration in growth was due to an increase in the use of prescription drugs in 2021.

3.2.3 Curb Excessive Spend—Analyze Markets

IQVIA, formerly Quintiles and IMS Health, Inc., is a multinational company serving the combined industries of health information technology and clinical research. It is a leading provider of biopharmaceutical development and commercial outsourcing services, focused primarily on Phase I-IV clinical trials and associated laboratory/analytical services, including consulting services. In its 2020 report on medicine spending, IQVIA researchers reported significant variations in medicine use, spending, and health outcomes globally. Overall, worldwide use of medicine from 2009 to 2019 has increased in areas of high priority, including treating noncommunicable diseases, which were responsible for 71% of deaths worldwide in 2016. This rise demonstrates the value of medicines in addressing the needs of patients, prescribers, and health systems worldwide (Kleinrock & Muñoz, 2020).

IQVIA also confirmed that medicine consumption is closely tied to country income, though there is significant variation across countries with similar incomes. The increase in medicine use is currently outpacing population and economic growth, indicating that more patients are receiving treatment. To assess these

Table 3.1 Ten leading sources of inefficiency in healthcare systems

Source of inefficiency	Common reasons for inefficiency	Ways to address inefficiency
1. Medicines: underuse of generics and higher than necessary prices for medicines	Inadequate controls on supply-chain agents, prescribers, and dispensers; lower perceived efficacy/safety of generic medicines; historical prescribing patterns and inefficient procurement/distribution systems; taxes and duties on medicines; excessive mark-ups	Improve prescribing guidance, information, training, and practice. Require, permit, or offer incentives for generic substitution. Develop active purchasing based on assessment of costs and benefits of alternatives. Ensure transparency in purchasing and tenders. Remove taxes and duties. Control excessive mark-ups. Monitor and publicize medicine prices
2. Medicines: use of substandard and counterfeit medicines	Inadequate pharmaceutical regulatory structures/ mechanisms; weak procurement systems.	Strengthen enforcement of quality standards in the manufacture of medicines; carry out product testing; enhance procurement systems with pre-qualification of suppliers
3. Medicines: inappropriate and ineffective use	Inappropriate prescriber incentives and unethical promotion practices; consumer demand/expectations; limited knowledge about therapeutic effects; inadequate regulatory frameworks	Separate prescribing and dispensing functions; regulate promotional activities; improve prescribing guidance, information, training, and practice; disseminate public information
4. Healthcare products and services: overuse or supply of equipment, investigations and procedures	Supplier-induced demand; fee-for-service payment mechanisms; fear of litigation (defensive medicine)	Reform incentive and payment structures (e.g., capitation or diagnosis-related group); develop and implement clinical guidelines
5. Health workers: inappropriate or costly staff mix, unmotivated workers	Conformity with pre-determined human resource policies and procedures; resistance by the medical profession; fixed/ inflexible contracts; inadequate salaries; recruitment based on favoritism	Undertake needs-based assessment and training; revise remuneration policies; introduce flexible contracts and/or performance-related pay; implement task-shifting and other ways of matching skills to needs
6. Healthcare services: inappropriate hospital admissions and length of stay	Lack of alternative care arrangements; insufficient incentives to discharge; limited knowledge of best practice	Provide alternative care (e.g., daycare); alter incentives to hospital providers; raise knowledge about efficient admission practice

(continued)

Table 3.1 (continued)

Source of inefficiency	Common reasons for inefficiency	Ways to address inefficiency
7. Healthcare services: inappropriate hospital size (low use of infrastructure)	Inappropriate managerial resources for coordination and control; too many hospitals and inpatient beds in some areas, not enough in others. Often, this reflects a lack of planning for health service infrastructure development	Incorporate inputs and output estimation into hospital planning; match managerial capacity to size; reduce excess capacity to raise occupancy rate to 80–90% (while controlling length of stay)
8. Healthcare services: medical errors and suboptimal quality of care	Insufficient knowledge or application of clinical-care standards and protocols; lack of guidelines; inadequate supervision	Improve hygiene standards in hospitals; provide more continuity of care; undertake more clinical audits; monitor hospital performance
9. Health system leakages: waste, corruption and fraud	Unclear resource allocation guidance; lack of transparency; poor accountability and governance mechanisms; low salaries	Improve regulation/governance, including strong sanction mechanisms; assess transparency/vulnerability to corruption; undertake public spending tracking surveys; promote codes of conduct
10. Health interventions: Inefficient mix/ inappropriate level of strategies	Funding high-cost, low-effect interventions when low-cost, high-impact options are unfunded. Inappropriate balance between levels of care and/or between prevention, promotion and treatment	Regular evaluation and incorporation into policy of evidence on the costs and impact of interventions, technologies, medicines, and policy options

Source: WHO

changes in volume, a daily defined dose[5] (DDD) metric was used to normalize use across countries, driving understanding of usage on a normalized day of therapy. DDD consumption reached 1.8 trillion DDDs in 2019, a 16% increase over 2014, corresponding to a 3% compound annual growth rate (CAGR). Pharmerging[6] markets saw the greatest CAGR of 4.4% and accounted for 49% of DDD consumption. National income positively correlated with medicine use, with richer countries typically having higher rates of medicine use, as shown in Fig. 3.7. While pharmerging countries are responsible for most DDD volume growth, many countries, such as

[5]WHO has developed a method of normalizing medicines of varying intended doses using a defined daily dose (WHO-DDD). The WHO-DDD *is the assumed average maintenance dose per day for a drug used for its main indication in adults.* The DDD is a unit of measurement and does not necessarily reflect the recommended or Prescribed Daily Dose. *The* IQVIA institute has developed their DDD factors using the same or highly similar concepts to represent over 99% of audited standard unit volume globally in their 2020 report (Kleinrock & Muñoz, 2020).

[6]Pharmerging markets are defined as countries with per capita income below $30,000 per year and pharmaceutical spending growth above $1 billion over 5 years.

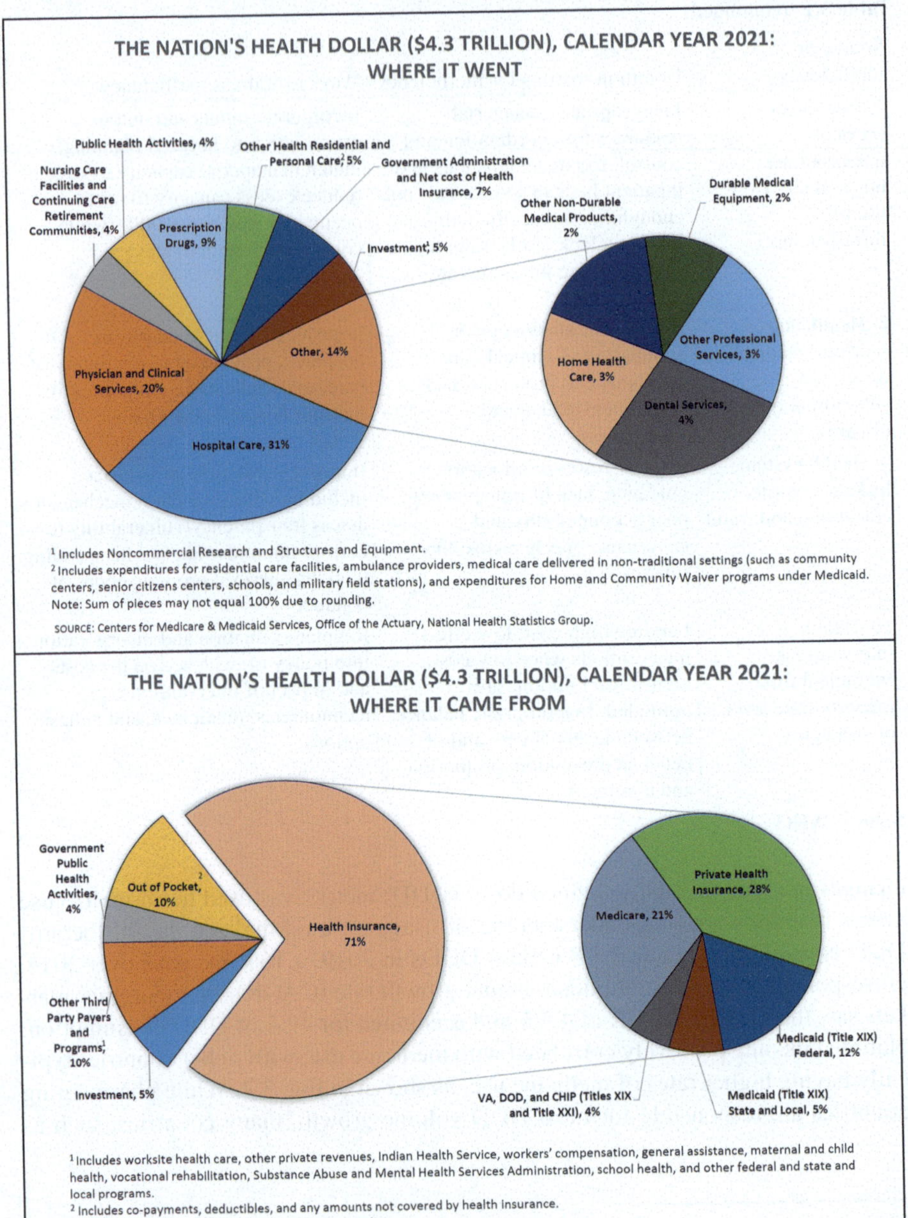

Fig. 3.6 Inputs and outputs of the US health market (CMS, 2021)

India, China, and Mexico, have low DDD per capita, indicating greater access and availability is still needed to reach more patients (Kleinrock & Muñoz, 2020).

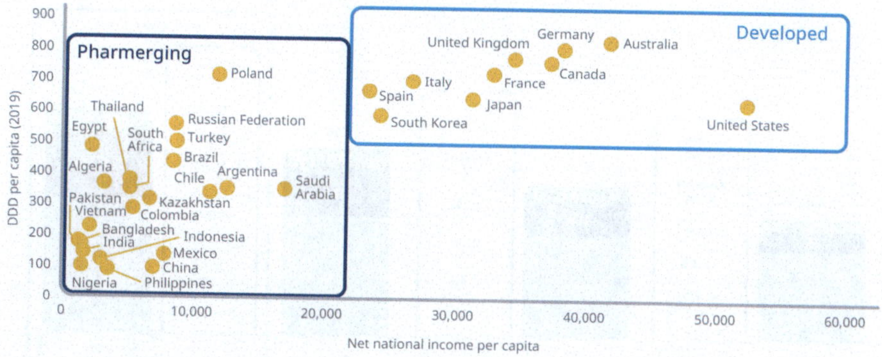

Fig. 3.7 Defined Daily Doses (DDD) vs Net National Income in 2019. (Source: Kleinrock & Muñoz, 2020)

WHO estimates that medicines account for 20–30% of global health spending, slightly more in low- and middle-income countries, and, therefore, constitute a major part of a country's health budget (WHO, 2010). IQVIA reports that global medicine spending growth is consistent with the increased global use of medicine and is expected to exceed $1.1 trillion in 2024, compared to $955 billion in 2019 and $777 billion in 2014. This rise in spending is occurring despite discounts/rebates and in all markets (Fig. 3.8). Other researchers have also looked into the trends for medicine spending, where they analyzed data from key international organizations, including the Organization for Economic Co-operation and Development (OECD).[7] Comparing underlying differences in structural features, types of healthcare and social spending, and performance between the United States and ten high-income countries, researchers found that in 2016, the United States had the highest pharmaceutical spending per capita at $1443, with Switzerland following at $939 and a mean of $749 for all 11 countries (Papanicolas et al., 2018).

The trend of high spending on pharmaceuticals is a global problem. Every 2 years, the OECD publishes a vital publication, *Health at a Glance*, which compares critical indicators for population health and health system performance across OECD member countries and key emerging economies. In its 2021 edition, *Health at a Glance* presents the latest comparable data, illustrating differences across countries and over time in terms of health status, risk factors for health, access to and quality of care, and health resources. Per the report, in 2019, spending on retail pharmaceuticals (excluding those used during hospital treatment) accounted for one-sixth of overall healthcare expenditure in OECD countries. It represented the third most significant component of health spending after inpatient and outpatient care. Across OECD countries, governments and compulsory insurance schemes accounted for the largest

[7] OECD is an intergovernmental organization with 38 member countries, founded in 1961 to stimulate economic progress and world trade. The majority of OECD members are high-income economies with a very high Human Development Index (HDI) and are regarded as developed countries. As of 2017, OECD member countries collectively comprised 62.2% of global nominal GDP (US$49.6 trillion).

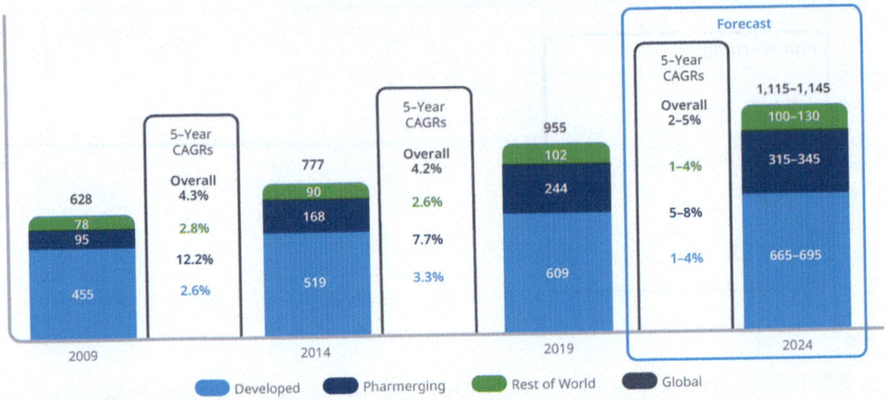

Fig. 3.8 Global medicine market size in constant US billion dollars (2009–2024). (Source: Kleinrock & Muñoz, 2020)

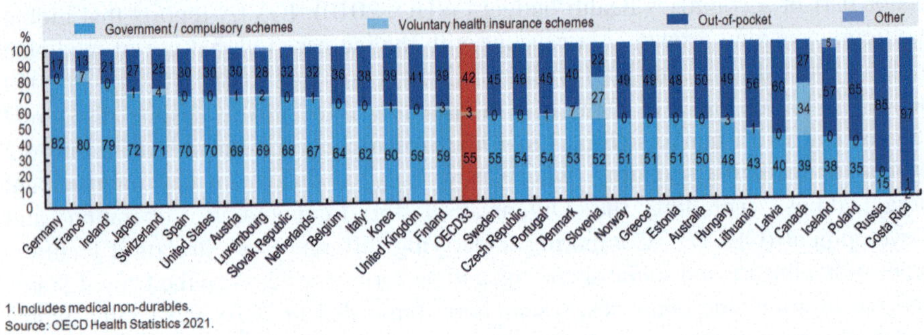

Fig. 3.9 Expenditure on retail pharmaceuticals by type of financing (2019 or nearest year) (OECD, 2021)

share of retail pharmaceutical costs, covering 56% of total spending (Fig. 3.9). This share was even higher in countries such as Germany and France, with more than 80% of total costs covered by these schemes. In contrast, voluntary health insurance schemes covered a relatively small proportion, averaging only around 3%. Among the exceptions were Slovenia and Canada, where almost one-third of pharmaceutical spending was covered by private insurance (OECD, 2021).

The other significant source of financing was out-of-pocket household payments (including cost-sharing for reimbursed drugs). Out-of-pocket spending amounted to an average of 41% of total pharmaceutical spending, albeit with much higher levels in countries such as Poland and Latvia, where it accounted for almost two-thirds of the total. Various factors influence the per capita spending on retail pharmaceuticals, including distribution, prescribing, and dispensing; pricing and procurement policies; and patterns of uptake of novel and generic medicines. In 2019, per capita retail pharmaceutical expenditure in OECD countries averaged $571 (adjusted for differences in purchasing power) (Fig. 3.10). Spending in the United States was more than double the OECD average, while most OECD countries fell within a

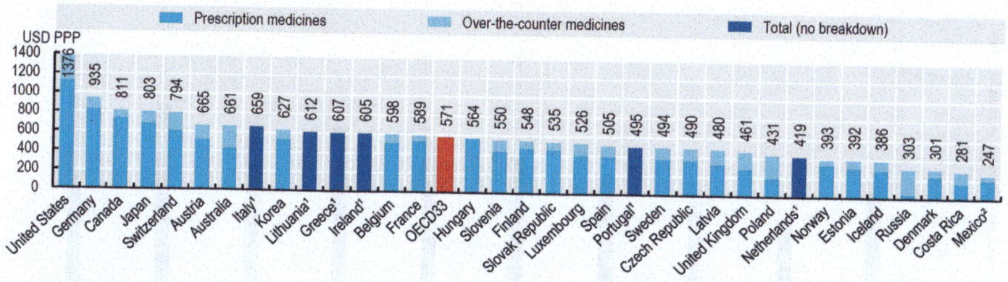

1. Includes medical non-durables (resulting in an overestimation of around 5-10%). 2. Only includes private expenditure.
Source: OECD Health Statistics 2021.

Fig. 3.10 Expenditure on retail pharmaceuticals per capita (2019 or nearest year) (OECD, 2021)

narrow spending band of ±15% from the average. Per capita spending was lowest in Mexico and Costa Rica, at less than half the OECD average (OECD, 2021).

Pharmaceutical expenditure has two main components: prescription medicines and over-the-counter (OTC) products. Across OECD countries in 2019, prescription medicines accounted for 79% of pharmaceutical spending (Fig. 3.10). Country-specific differences influence the prices and availability of different prescription medicines. For example, in the United States, per capita spending on prescription medicines is ~$1200, the highest in the world. Poland was the only OECD member country where spending on OTC products exceeded that on prescription medicines. In the United Kingdom and Australia, OTCs accounted for a third of total pharmaceutical expenditure, while in Canada and France, spending on prescription medicines accounted for 90% of the total.

In the specific case of the United States, a vital agency of the Department of Commerce is the Bureau of Economic Analysis (BEA). BEA produces economic account statistics that enable government and business decision-makers, researchers, and the American public to follow and understand the performance of the nation's economy. BEA also produces the foremost measure of spending by US consumers, also called personal consumption expenditures (PCE). PCEs include how much is spent on durable (such as motor vehicles, furnishings, recreational goods), nondurable goods (such as food, clothing, gasoline), and services.[8] It accounts for about two-thirds of final domestic spending, and thus, it is the primary engine that drives future economic growth. With the information on consumer spending, we can evaluate overall spending and its rate of change and predict future trends. Consumer spending is a key component of the quarterly gross domestic product. BEA's monthly consumer spending statistics also are closely watched as early gauges of the economy's strength (BEA, 2022). The data from BEA on the PCE from Q1 2020 to Q2 2022 is plotted in Fig. 3.11. The data brings us a few important questions to ponder. Why is the spending on prescription medicine so high in the United States? What can be learned from other countries where prescription medicine spending is proportionally less than in the OTC? It is critical to evaluate how pharmaceutical pricing policies are determined to answer these questions.

[8] A service is anything that a business provides to an individual that they can't do for themselves. Healthcare falls within the "Services" category.

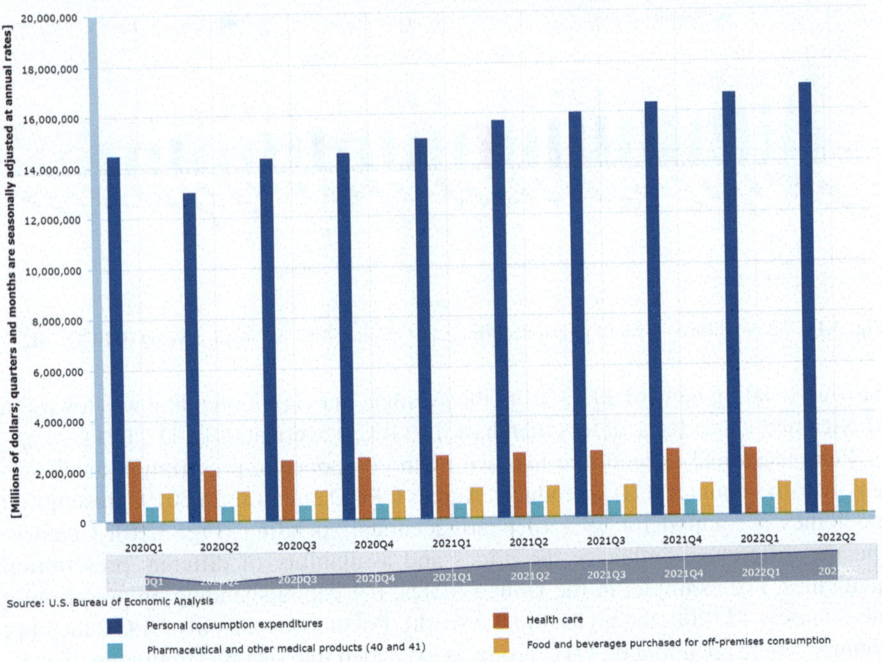

Fig. 3.11 PCE expenditures (Q1 2020–Q2 2022) in the United States (BEA, 2022)

Why is the spending on prescription medicine so high in the United States?

3.2.4 Regulate Pharmaceutical Pricing

Pharmaceutical pricing is a complex topic but an important one. It is incredibly complex in the United States because, unlike other developed countries, the United States does not negotiate over the price of a new drug based on the value it provides. The lack of value-based pricing (VBP) in the United States also has a direct adverse effect on the ability of other countries to negotiate prices with manufacturers. It dramatically reduces the leverage that individual countries have. Also, in addition to not having a system for VBP, the United States has specific legislation that prohibits the biggest purchaser of oral prescription drugs, Medicare, from directly negotiating with manufacturers (Vincent Rajkumar, 2020) (*please see* Sect. 8.2 *for a detailed discussion on the 2022 Inflation Reduction Act that aims to reduce the cost of certain prescription drugs covered by Medicare*). The result is that in the United States, there is no single price for a drug that all participants know. Instead, drug pricing has many different prices depending on who is buying, who is selling, and when and where the transaction occurs. These multiple prices correspond broadly to different levels of the distribution chain and whether the price is a list price or a transaction price (that reflects discounts or rebates off the list price) (Danzon, 2014). The range of prices paid in the market helps identify the many players beyond the manufacturer and final consumer. These players are the wholesalers, retailers, and pharmacies.

- ***Wholesalers:*** The major purchasers of drugs from manufacturers are not patients or pharmacies but wholesalers such as AmerisourceBergen, Cardinal Health, etc. Wholesalers account for 85% to 90% of drug manufacturer revenues and purchase drugs directly from the manufacturers for sale to pharmacies, hospitals, physician offices, and stand-alone clinics (AHP, 2022). The price wholesalers pay to purchase drugs from manufacturers is called the Average Manufacturer Price (AMP) or Wholesale Acquisition Cost (WAC), sometimes with modest discounts (1–2%) for prompt payment. Medicaid and some other public payers, by statute, get mandatory rebates from AMP. (Danzon, 2014).
- ***Retailers:*** The next stage in the drug distribution pipeline is the sale of pharmaceuticals from wholesalers to retailers. Focusing on pharmacies that sell directly to consumers, retailers' price is often known as the Actual Acquisition Cost (AAC). The AAC is typically based on the WAC plus a markup (usually 10–15% on branded drugs and higher on generics). Average Wholesale Price (AWP) is another benchmark for the price pharmacies pay wholesalers. AWP is a universal standard in pharmaceutical pricing and is typically collected and published by companies that collate drug pricing data. This list price was widely used by payers as a basis for their reimbursement to pharmacies. However, AWP became an increasingly unreliable (usually inflated) measure of the actual average transaction price at which wholesalers sell to pharmacies. Although AWP and other list prices have remained the basis for payer reimbursement to pharmacies, there is usually a significant discount. For example, a payer may set pharmacy reimbursement at an 18% discount off AWP, where the discount is negotiated between the payer and the pharmacy chain. This reimbursement price is intended to allow the pharmacy to cover the cost of drug acquisition plus a competitive dispensing fee. The payer reimburses the pharmacy at this price, net of any patient cost-sharing that the pharmacy must collect from the patient, depending on the payer's plan (Danzon, 2014).
- ***Pharmacies:*** The last step is getting medications into the hands of consumers. This is handled through several chains, local retail pharmacies, and an increasing number of mail and specialty pharmacies. The retail pharmacy market in the United States is primarily dominated by chain pharmacies such as Walgreens, CVS Health, and Rite Aid. An increasing volume of drugs is being dispensed through the mail order channel, especially with the expansion of specialty drug utilization. The price of retail medications to consumers is the "Usual and Customary" (U&C) price, which includes the cost of the drug (AAC) plus the pharmacy's markup. The pharmacy typically also receives a $1–$3 dispensing fee per prescription. Figure 3.12 shows the manufacturers' and final consumers' various prices (AHP, 2022).

An essential role in drug purchasing and pharmacy pricing is played by insurers and pharmacy benefit managers (PBMs). Typically, consumers with pharmacy insurance pay a copay or a percentage of a drug's cost, and their insurance covers the remainder. The pharmacy costs covered by insurance are often lower than other

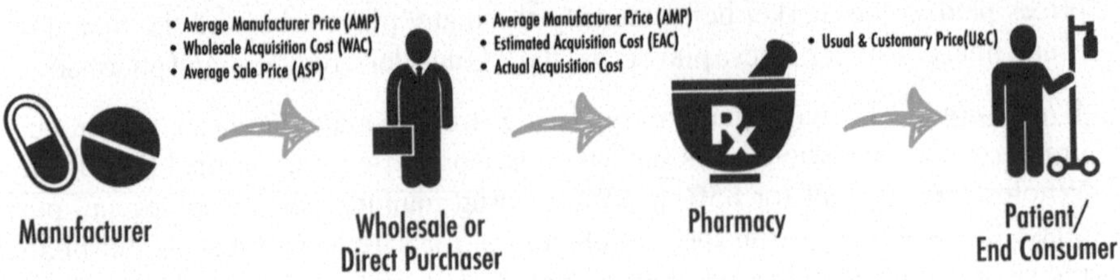

Fig. 3.12 Lifecycle of pharmaceutical pricing (AHP, 2022)

medical services, but they have risen recently, especially for costly specialty medications. Insurers entered the pharmaceutical market to use their market power to reduce the prices they pay for drugs. Over time, many insurers have outsourced this role to PBMs, which negotiate drug prices for insurers and large employers. PBMs work on behalf of their clients to lower the prices paid for pharmaceuticals. They interact in the pharmaceutical market through two primary paths: price negotiation and formulary design. The first part of price negotiation is reducing the prices paid at the pharmacy through discounts. PBMs aggregate the purchasing power of multiple insurers and payers to negotiate better discounts with pharmacies than insurers could achieve on their own (AHP, 2022). All of this results in a US health insurance system structured so that multiple intermediaries modulate patient care and accessibility of prescribed medications. This system creates a complex pharmaceutical supply chain that clouds the financial transparency among the transactions, as shown in Fig. 3.13. The net result is that the accurate prices of medications are difficult to decipher because of the opaque contracts and incentives that insurers, drug makers, pharmacies, distributors, and PBMs use to exchange money (Psotka et al., 2020). In order to understand this aspect of healthcare, it is equally necessary to consider the perspectives of a significant stakeholder in healthcare systems—the insurance companies.

3.3 Insurance Companies—Manage Healthcare Market Efficiently

From an economic perspective, health is a unique commodity different from all other tangible commodities and assets, such as food, shelter, clothing, and other material possessions. In a classical health economics model developed in the 1970s, everyone is both a producer and a consumer of health. It is assumed that individuals inherit an initial health-stock that depreciates over time—at an increasing rate, at least after some stage in the life cycle—and can be increased by investment. Direct inputs into the production of gross investments in the health stock include own time, medical care, diet, exercise, housing, and other market goods. The production function also depends on certain "environmental variables", the most important of which is the level of the producer's education, which alters the production process'

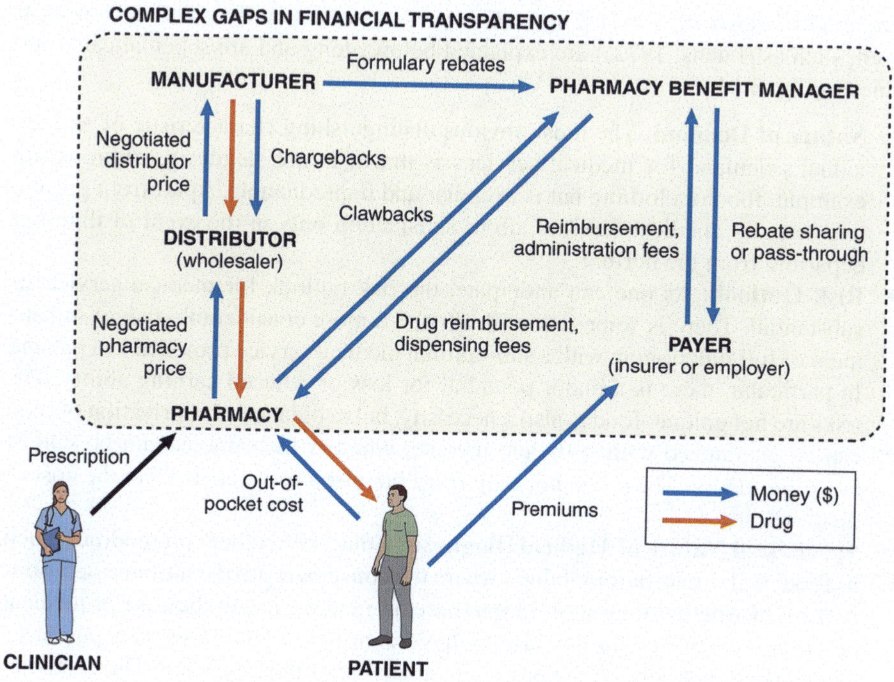

Fig. 3.13 Complex gaps in financial transparency among transactions in the US pharmaceutical supply chain (Psotka et al., 2020) (reprinted with permission from Wolters Kluwer Health, Inc.)

efficiency (Grossman, 1972). This fundamental model allows healthcare economics to be parameterized, and its marketplace dynamics can be better understood.

3.3.1 *Dynamics of the Healthcare Market*

In healthcare, innovative solutions that help make a significant societal impact and are affordable and profitable must be developed to justify continued investment in this area while operating under strict regulatory oversight and guidance. So, how is healthcare as a commodity different from other commodities? In a seminal journal article from the 1960s, Professor K.J. Arrow, a mathematician, writer, political theorist, and Nobel Laureate in Economic Sciences, provides a treatise on the uncertainty and the welfare economics of medical care.[9] Dr. Arrow selectively lists some characteristics of medical care that distinguish it from the usual economics of other

[9] In his article, Dr. Arrow differentiates between healthcare industry and the medical-care industry. As per Dr. Arrow, the causal factors in health are many, and the provision of medical care is only one. In the case of general healthcare, particularly at low levels of income, other commodities such as nutrition, shelter, clothing, and sanitation may be much more significant. On the other hand,

commodities (Arrow, 1963). These characteristics (collated by other references as well, such as (Fuchs, 1972)) are explained below along and are schematically outlined in Fig. 3.14:

- **Nature of Demand**: The most obvious distinguishing characteristic of an individual's demand for medical services is that it is not steady in origin as, for example, food or clothing but is irregular and unpredictable. Apart from preventive services, medical services afford satisfaction only in the event of illness, a departure from the normal.
- **Risk Outlook**: As one can anticipate, the risk outlook for medical services is substantial. There is some risk of death and a more considerable risk of impairment of full functioning with a sub-optimal medical service provided to a patient. In particular, there is a major potential for loss or reduced earning ability. The risks are not unique; food is also a necessity, but avoidance of deprivation of food can be guaranteed with sufficient income, whereas the same cannot be said of avoiding illness. Thus, it is not only risky but also costly, apart from the cost of medical care.
- **Specialized Nature of Medical Diagnosis**: Contrary to other commodities (such as food, real estate, automobiles), where the consumers are the ultimate decision-makers of whether they wish to purchase an item or not, in the case of medical services, consumers do not always have control on what they can purchase. Consumers must rely on the professional advice of the physician. The customer cannot test the product before consuming it, and there is an element of trust in the relationship. Therefore, medical services have a specialized operating model that is significantly different from other typical commodities.
- **Difficulties in Innovation**: As discussed earlier, the human body is complex in structure and functionality. To provide cures and relief to patients, it is increasingly vital for the medical community and pharmaceutical industry to innovate continuously. Continued research in this area is a win-win scenario for society. However, a favorable outcome is not always guaranteed from an economic perspective. Substantial financial risk must be undertaken without any assurance of economic reward. This type of significant financial investment is very different from the investments made in other commodities.
- **Pricing and Reimbursement**: Pricing and reimbursement (P&R) involve negotiations between manufacturers and P&R authorities regarding the price of the new product and its reimbursement status. Price regulation can arguably delay launch through the negotiation processes, alongside the resulting firm strategies of delaying or foregoing launch in low-priced markets (Varol et al., 2012).

As one examines Fig. 3.14, it is clear that insurance companies are essential stakeholders in patient care. In addition to helping people stay healthy and improving their health when they get sick, the health insurance system serves important functions to enable everyone to benefit from being insured (Dey & Bach, 2019).

---

medical-care industry is defined as the complex of services that center about the physician, private and group practice, hospitals, and public.

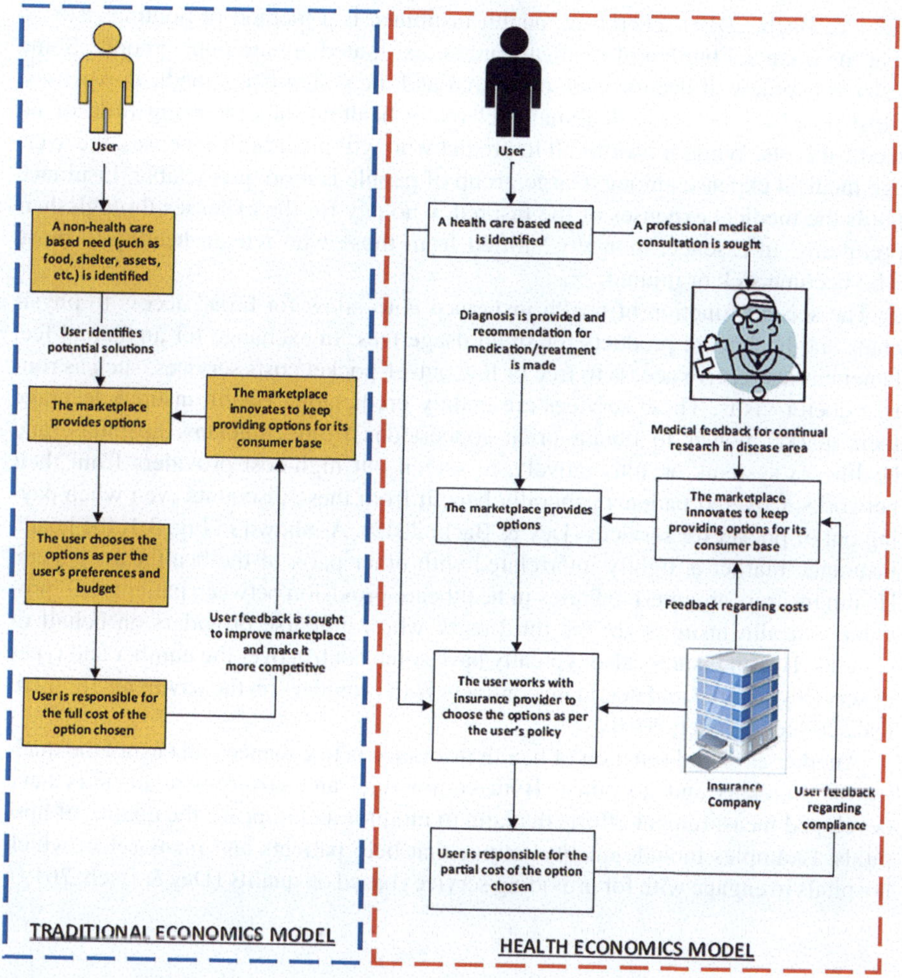

Fig. 3.14 Traditional vs. health economics model (Mittal, 2019) (reprinted with permission from Elsevier)

3.3.2 *Functions of Health Insurance*

The first and foremost function of the health insurance system is to protect individuals and families financially during a catastrophic (and unpredictable) health event. People typically buy insurance to avoid risks by paying a known premium to receive benefits if an adverse event occurs during the insurance policy's term. Most individuals are willing to pay an insurer to assume the bulk of financial risks associated with unpredictable health outcomes of uncertain severity. Policies to improve this function of health insurance include capping annual out-of-pocket expenses, ending lifetime benefit limits, and ensuring coverage for people with preexisting conditions

(Dey & Bach, 2019). Therefore, health insurance is a method of pooling risks so that the financial burden of medical care is distributed among many people. Some insured people will become sick or injured and incur significant medical expenses. Most people, however, will remain relatively healthy, thus incurring little or no medical costs. While it is difficult to predict who will incur high expenses, the average medical expense among a large group of people is more predictable. Insurance pools the medical expenses of the insured, who pay for the expenses through their premiums. In essence, money is shifted from those who remain healthy to those who become sick or injured.

The second function of health insurance is to allow for broad access to physicians and healthcare products for small usage fees. In exchange for an annual fee, beneficiaries receive access to free or low out-of-pocket costs services, such as routine doctor visits. These services are mainly predictable. Health insurers leverage their market power to obtain price concessions from clinicians, hospitals, and healthcare systems or, alternatively, to screen out high-cost providers from their networks. Covered patients generally benefit from these discounts even when paying out of pocket for services (Dey & Bach, 2019). As shown in Fig. 3.1, the health insurance market is tightly interrelated with other parts of the healthcare system. Health insurers are intermediaries in healthcare provision between patients and providers—health insurers are the third party who reimburse providers on behalf of patients. Health insurers also typically have some control over the number and types of services covered and negotiate contracts with providers on the payments for medical services (Jenner, 2010).

Another essential function of health insurance is to enhance and ensure the quality of clinicians and hospitals. Both commercial and government insurers have developed measurement efforts that aim to monitor and improve the quality of hospitals. Examples include quality ratings that help patients and plans select which hospitals to engage with for providing services based on quality (Dey & Bach, 2019).

3.3.3 *Administration of the Health Insurance Market*

According to economic theory, perfectly competitive markets allocate goods and services efficiently if certain conditions are met. Markets allocate goods and services efficiently when the social cost of the resources (e.g., labor, buildings, machinery, raw materials) used to make the last unit sold equals the social benefit of consuming that unit. Conditions required to ensure the efficiency of competitive markets include the presence of many buyers and sellers, low barriers of entry for new firms, fairness in competition, symmetric information,[10] and no unnecessary

[10] Symmetric information in economics is one of the main requirements for the efficient functioning of competitive markets, and it helps to prevent sellers from extracting an excessive price from buyers if those buyers have an equal amount of information about a given product. Most of economic science is built upon the simplifying assumption that all knowledge is symmetric and com-

transaction costs. Unfortunately, the health insurance market has many features that push it far from the economic benchmark of perfect competition.

This leads to inefficient allocation of services and resources, thereby creating health inequalities. For example, in the current US healthcare market, how insurers design healthcare networks influences how consumers use healthcare. Consumers typically choose a primary physician who selects tests and treatments and makes referrals to medical specialists. Employers negotiate with insurers on behalf of their workers, and labor unions negotiate with employers over health benefits for their members. Health insurers, in turn, negotiate contracts with providers and handle payments for individual services. A primary physician's admitting privileges typically determine where his patient goes for non-emergency hospital care. Patients must go through a physician to obtain most medical tests and pharmaceuticals. Healthcare consumers typically rely on these intermediaries instead of interacting directly with other parts of the healthcare system. This creates a system of heavy reliance on intermediaries that protect consumers from financial risks and insulate them from information about costs and prices for specific healthcare goods and services (Jenner, 2010).

In the United States, most health insurance plans are managed care plans (HMOs or PPOs) rather than indemnity or traditional health insurance plans that provide unlimited reimbursement for a fixed premium. A Health Maintenance Organization (HMO) is a network of doctors, hospitals, and other healthcare providers who agree to provide care at a reduced rate. To keep costs low, HMOs may require their subscribers to select a primary care physician (PCP) who refers the patient to specialists when needed. Like an HMO, a Preferred Provider Organization (PPO) is a network of doctors, hospitals, and healthcare providers who agree to provide care at a specific rate. Unlike an HMO, subscribers of PPO are typically not limited to providers in the network, thereby providing increased flexibility, though the copay or out-of-pocket cost for out-of-network visits may be higher than for in-network providers.

Other parties involved in the health insurance market include employers (most private health insurance is obtained through an employer), federal, state, and local governments, and healthcare providers. At the federal level, the Centers for Medicare and Medicaid Services (CMS) administers both Medicare and Medicaid, the fundamental pillars of the US government's healthcare policy. Medicare is financed by payroll taxes, known as the Federal Insurance Contributions Act (FICA), which equals 2.9% of income, with 1.45% withheld from the worker and a matching 1.45% paid by the employer. The self-employed must pay the entire 2.9% tax on self-employed net earnings. Medicare contracts with regional insurance companies to pay doctors and hospitals for medical services. In the case of Medicaid, it is jointly financed by the federal and state governments. No taxes are directly collected for Medicaid. The federal government pays a matching rate of funds to what states spend for the plan, but that amount varies from state to state and is related to

plete i.e., that both buyers and sellers have full information about the products on the market. The benefits of symmetrical information in business is to allow the efficient functioning of the price mechanism.

a state's per capita income. It is usually 57% of the costs. State participation in Medicaid is optional, but every state does participate. Some states require local communities to contribute toward the financing. In many states, Medicaid beneficiaries must pay fees or co-payments for medical services. Like Medicare, Medicaid does not pay benefits directly to individuals. Instead, it sends benefit payments to healthcare providers (Koba, 2011).

CMS administers the Medicare program and works in partnership with state governments to administer Medicaid, the Children's Health Insurance Program (CHIP), and health insurance portability standards. In addition to these programs, CMS has other responsibilities, including the administrative simplification standards from the Health Insurance Portability and Accountability Act of 1996 (HIPAA), quality standards in long-term care facilities through its survey and certification process, clinical laboratory quality standards under the Clinical Laboratory Improvement Amendments, and oversight of the website, HealthCare.gov. A substantially large number of people are enrolled in the programs administered by CMS (CMS, 2019) (Table 3.2).

3.3.4 Improve Accessibility to Medications

One of the primary goals of having health insurance is to improve access to medications. However, as shown in Fig. 3.13, the US health insurance system creates a complex pharmaceutical supply chain that clouds the financial transparency among

Table 3.2 Programs administered by CMS

Program	Details
Medicare[a]	Medicare is a single-payer, national social insurance program for people aged 65 or older, people under age 65 with specific disabilities, and people of all ages with end-stage renal disease (permanent kidney failure requiring dialysis or a kidney transplant). Medicare has the following parts: Part A hospital insurance - helps cover inpatient care in hospitals, including critical access hospitals and skilled nursing facilities (not custodial or long-term care) Part B medical insurance - helps cover doctors' services and outpatient care. It also covers some other medical services that part A does not cover Part D prescription drug coverage - Medicare prescription drug coverage is insurance that helps lower prescription drug costs and protects against higher prices in the future
Medicaid	Medicaid is a joint federal and state program that helps with medical costs for some people with limited income and resources. Medicaid offers benefits not usually covered by Medicare, like nursing home care and personal care services
CHIP	CHIP is a program that provides matching funds to states for health insurance for families with children. The program was designed to cover uninsured children in families with modest incomes too high to qualify for Medicaid

[a]Medicare Advantage plans (Part C) are run by private insurance companies approved and subsidized by Medicare. They must cover all services that the original Medicare covers except hospice care, but they can offer extra coverage, sometimes at additional cost, for vision, hearing, and dental, and they may have different rules regarding how enrollees receive services (Mazie, 2024)

the transactions. In contrast, many non-US health insurance systems are governmental programs that control therapeutic use, either as part of universal coverage or as a tightly regulated multi-payer arrangement. In Britain, for instance, the National Health Service typically provides only medications and technologies recommended by the National Institute for Health and Care Excellence, which both function as the national regulatory authority and appraise the cost-effectiveness of interventions (Psotka et al., 2020).

One way the US system increases access to medication for pharmaceutical classes for which effective generic competition exists (such as statins) is by successfully moving patients to the cost-saving generic alternatives. These actions can protect patients from ineffective, inappropriate, or harmful prescriptions, for instance, by encouraging generic medications if newer and more expensive options lack evidence of superior safety or efficacy. However, it also creates a scenario where insurance companies can actively restrict some medications to medically accepted indications defined by FDA labeling or as delineated in evidence-based compendia (Psotka et al., 2020). Access to novel therapeutics is a distinguishing feature of the US health system and a key driver of high spending. Highly effective therapeutics may lower aggregate spending by reducing the need for costly interventions or hospitalizations, but many of these novel medicines command high prices. More challenging is that high prices do not always align with value. Prices of existing, branded drugs have increased substantially during the past decade, limiting affordability and access (Shrank et al., 2021). This brings us to a fundamental issue again. ***Why are prescription medicines so expensive*?**

3.4 Innovative Pharmaceutical Industry's Perspective—R&D Is Expensive

US' global pharmaceutical leadership is built upon an industry that performs and supports advanced R&D, resulting in innovative new treatments and cures. Novel drugs serve previously unmet medical needs or significantly help advance patient therapies. The active ingredient(s) in a novel drug have never been approved. Companies that deal with developing innovative products spend significant resources on drug discovery and development. As part of drug discovery, innovative pharmaceutical companies focus on understanding the metabolic pathways related to a disease state (or pathogen) and manipulate these pathways using molecular biology or biochemistry. Hundreds and sometimes thousands of chemical compounds must be synthesized and tested to find one that can achieve a desirable result. Once a potential drug compound is identified, product development activities are undertaken to establish its safety and efficacy through a combination of *in vitro* studies, *in vivo* studies, and clinical trials. Under FDA requirements, a sponsor must first submit data showing that the drug is reasonably safe for initial, small-scale clinical studies. This process is achieved through the Investigational New Drug (IND) pathway.

As the clinical studies start to yield favorable data, the sponsor may choose to pursue the drug's commercialization path in the United States using either the New Drug Application (NDA) pathway for New Chemical Entities (NCE) or the Biologics License Application (BLA) pathway for biologics. The product could only be marketed in the United States after the FDA approved the NDA/BLA.

Although estimates vary by indication, the US Food and Drug Administration (US-FDA) assesses that it takes approximately eight-and-a-half years to study and test a new drug before it can be approved. This estimate includes early laboratory and animal testing and later clinical trials using human subjects (FDA, 1998). During the entire time, millions of dollars were expended to develop the drug, along with thousands of experiments done by the various disciplines involved to prove the dosage form's safety, efficacy, manufacturability, and quality. However, with all the time, resources, and money spent on research, the unfortunate part is that favorable results are never guaranteed.

3.4.1 Business Proposition of Pharmaceuticals

From a business perspective, the pharmaceutical industry has a profitable but precarious operating model. Two types of risks are involved in developing and commercializing a new drug: technical and market risks. Technical risks are associated with uncertainties as to the technological aspects of drugs. Market risks and uncertainties are related to commercial potentials, which are affected by the size of the target disease's patient population, new competing drugs entering the market, third-party payer's payment regulations, and the economy in general (Park & Shin, 2018). The successful drugs in the marketplace are typically the only sources of revenue for a company. These successful drugs also pay for the failed drugs. As expected, in any company, R&D costs are paid for by drugs that bring in revenues. Ideally, the cost of drug development should be less than the amount of money that will be made on the drug. This can be displayed in a hypothetical cash flow diagram (Fig. 3.15). The expenses incurred during R&D are displayed as negative cash flow, and revenues from the drug's sales create positive cash flow. Until the product launch, the cash flow is negative due to R&D expenses. Upon launch, the drug sales bring in revenues, which are (hopefully) greater than the expenses, and cash flow becomes positive. However, cash flow is severely impacted when the drug's exclusivity is lost, allowing generics to enter the market (Samanen, 2013).

3.4.2 Hurdles in Pharmaceutical Innovation

Pharmaceutical research is complicated, time-consuming, and costly. The unfortunate part is that the result is never guaranteed. To further complicate matters, there is no standard route through which drugs are developed. A pharmaceutical company

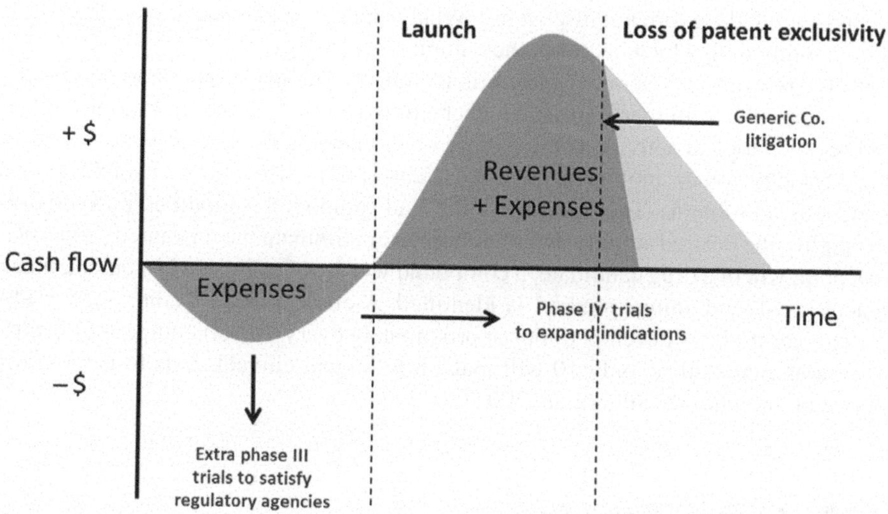

Fig. 3.15 Hypothetical cash flow over time and milestones for pharmaceutical companies. (Samanen, 2013) (Reprinted with permission from Elsevier)

may decide to develop a new drug aimed at a specific disease or medical condition. Sometimes, scientists choose to pursue an exciting or promising line of research. In other cases, new findings from universities, government laboratories, or other laboratories may point the way for drug companies to follow with their research. Typically, no matter how a drug is developed, the general challenges from conception to commercialization remain the same. Typically, companies facing a competitive environment would like to launch new products as quickly as possible into several markets while the product is still under patent protection to amortize the substantial R&D outlays. However, pharmaceutical innovation has a few practical hurdles, as discussed below.

The human body is complex. Many diseases, or at least the symptoms, arise from an imbalance (either excess or deficiency) of particular chemicals in the body, invasion of a foreign organism, or aberrant cell growth. The effects of the imbalance can be corrected by antagonism or agonism of a receptor or by inhibiting a particular enzyme; foreign organism enzyme inhibition or interference with DNA biosynthesis or function are essential approaches to treat diseases arising from microorganisms and aberrant cell growth. A medicine can change a disease's course, alter an organ's function, relieve symptoms, or ease pain. Drugs come from a variety of sources, including plants, animals, and microorganisms. For example, of the 20 leading drugs in 1999, only 9 were derived from natural products. Also, almost 40% of the 520 new drugs approved for the drug market between 1983 and 1994 were natural products or derived from natural products (Silverman, 2004). Many modern medicines are synthetic versions of substances found in nature. However, sometimes drugs are entirely new chemicals that are not versions of natural substances.

In general, drugs are not discovered. What is more likely discovered is known as a lead compound, which becomes the natural receptor ligands or enzyme substrates once the relevant biochemical system is identified. The lead compound is a prototype compound with many attractive characteristics, such as the desired biological or pharmacological activity, but may have other undesirable characteristics, such as high toxicity, other biological activities, absorption difficulties, insolubility, or metabolism problems. The structure of the lead compound is modified by synthesis to amplify the desired activity and to minimize or eliminate the unwanted properties to a point where a drug candidate, a compound worthy of extensive biological, pharmacological, and animal studies, is identified. A clinical drug, a compound ready for clinical trials, is developed. For approximately every 10,000 compounds evaluated in animal studies, only 10 will make it to human clinical trials to get 1 compound on the market (Silverman, 2004).

3.4.3 *Limited Market Size*

An important metric for innovative products is the projection of the potential market size. As expected, market share directly correlates with financial investment and returns. If the anticipated market share is large, a larger financial investment can be justified for a higher rate of return. Pharmaceuticals are no different in this regard. The for-profit companies in the pharmaceutical industry choose innovation projects that they expect to be profitable. Profits will be determined by costs and revenues, which depend on the market size. The expected market share is broadly influenced by three factors (Dubois et al., 2015).

1. The first factor considers issues such as demographic and socioeconomic change, which affect the number of people likely to suffer from a particular medical condition and the resources they are likely to have available to spend on alleviating their condition.
2. The second factor is particular to the pharmaceutical and healthcare industries, such as the degree of competition among companies and the strategies that companies use to innovate, cut costs, and win customers, which affect the profitability of innovation.
3. The third factor concerns public policies, including policies toward intellectual property protection, drug safety, testing, pricing, reimbursement, and public research funding.

Different therapeutic markets whose potential sizes change differently due to demographic and income differences drive commensurate changes in innovative activity. For example, firms will want to serve that demand if the cardiovascular market is expected to grow due to an aging population. Therefore, there will be an increase in cardiovascular R&D before the demand materializes, and new cardiovascular products will vie for approval and marketing authorization. Based on the collated information from the top 25 pharmaceutical companies' current research areas, an

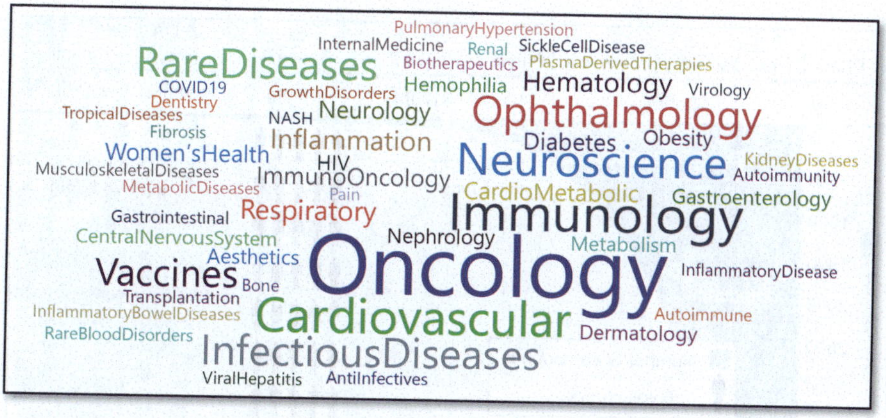

Fig. 3.16 Research areas in top 25 innovative pharmaceutical companies

interesting word cloud emerges that clearly shows the various research areas the companies are working on (Fig. 3.16).

3.4.4 *Extensive Clinical Trials*

No matter how medicine is made, its effect on the human body is far from simple. Drugs differ in how long they stay in the body, how easily they can get into different parts of the body, and how they are absorbed and eliminated by the body. People of different ages, genders, weights, and in different states of health respond differently to the same drug. These circumstances can alter how a drug is broken down and processed in the body. For example, elderly patients can respond differently to drugs because their kidneys eliminate drugs less effectively, and their liver breaks down drugs less efficiently. Similarly, when developing drugs for children, it is critical to recognize that their immature organ systems process differently than their mature bodies will in the years ahead.

The drug development process consists of three main stages: drug discovery, preclinical development, and clinical trials. Drug discovery starts with the finding of a hit molecule. A hit is a molecule that elicits a desired activity in a screening assay. Then, the structure of this molecule is optimized in terms of improving affinity and selectivity, reducing toxicity, improving water and lipid solubility, improving Absorption, Distribution, Metabolism and Excretion (ADME) properties in general, and converting the hit molecule into a lead molecule. The further optimization of the lead molecule delivers the drug candidate. Next, the preclinical studies are focused on clarifying the mode of action of the drug candidate and its pharmacokinetics in animals, such as bioavailability, toxic metabolites, if any, routes of excretion, efficacy in animals, drug formulation, and stability. Once a compound is selected for human clinical trials in the United States, a company must submit an IND

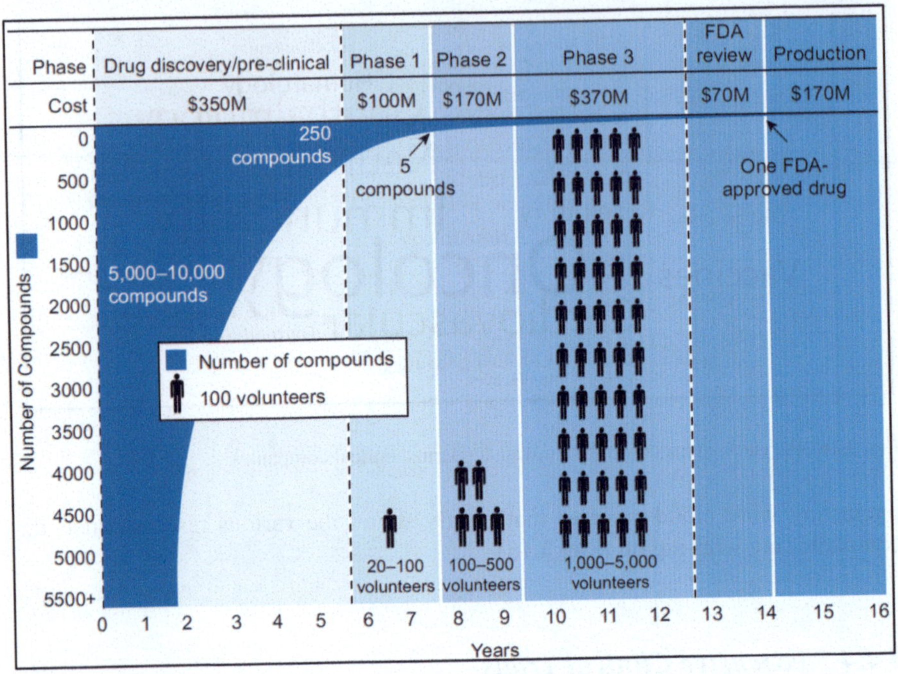

Fig. 3.17 Drug development process and attrition rates. (Burrill, 2014) (Reprinted with permission from Elsevier)

application to the FDA documenting what is known about the molecule, its activity, and any known safety concerns, as well as plans for how the clinical trial will be conducted and what they will seek to demonstrate. During this process, researchers must determine how a molecule would be delivered to a patient, what dosing would be used, and whether it could cause unintended effects.

Companies must overcome at least two regulatory hurdles before commercializing a new drug product. If approved for human clinical testing, the molecule will be subjected to progressively larger and more expensive human clinical trials to establish a particular drug's safety, dosing, and efficacy to treat a specific disease, as shown in Fig. 3.17 (Burrill, 2014). In this first hurdle, manufacturers must prove the three-fold requirement of quality, safety, and efficacy of new molecules, a process which is estimated to require around 10–15 years of pre-clinical and clinical research (Fig. 3.18). If the clinical testing is successful, the drug developer submits an NDA or BLA to the FDA—or a similar application to the regulatory authority in the country where it hopes to market the drug. The second hurdle typically includes the review of the new product dossier by the regulatory authority, such as the FDA, to get approval for marketing authorization. The application includes detailed data from the clinical trials and must provide sufficient information, data, and analyses to permit FDA reviewers to reach several key decisions, including:

- *Is the drug safe and effective for its proposed use(s), and does its benefits outweigh its risks?*

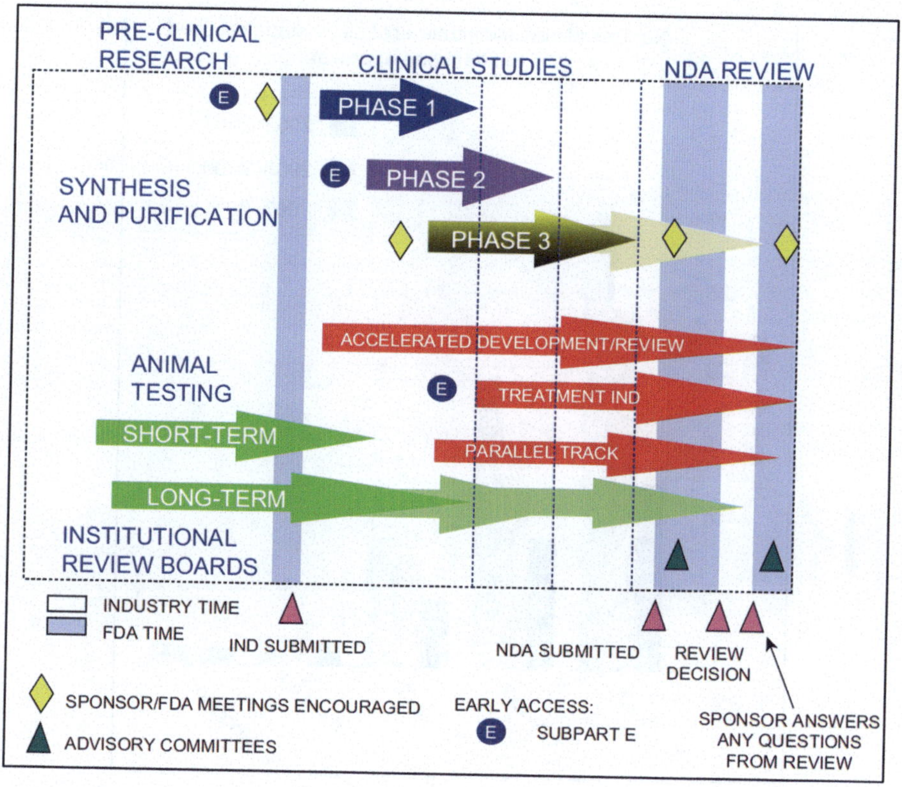

Fig. 3.18 New drug development process. (Source: FDA)

- *Is the drug's proposed labeling appropriate, and if not, what should the label contain?*
- *Are the methods used in manufacturing the drug and the controls used to maintain its quality adequate to preserve its identity, strength, quality, and purity?*

In most cases, the FDA is expected to respond within 10 months of the filing (Burrill, 2014). As one can anticipate, anything that can be done to shorten this development timeline can help bring therapies and medicines faster to the patients and help alleviate suffering. Unfortunately, very few new drugs reach approval. The statistics of new molecular entities (NMEs) going through various clinical stages to attain one approval are sobering, as shown in Fig. 3.19. This low R&D productivity necessitates higher and higher out-of-pocket costs for drug development. Researchers estimate that out-of-pocket costs for drug development rose nearly sixfold from the 1970s to the 2000s. The rising costs, in part, are due to falling success rates for drugs in clinical development as drugmakers sought to tackle more intractable diseases, such as Alzheimer's and cancer. Researchers found that success rates fell to 1 in 10 in the 2000s compared to 1 in 5 during the 1980s. The increasing complexity of the science underlying drug development and rising regulatory barriers have

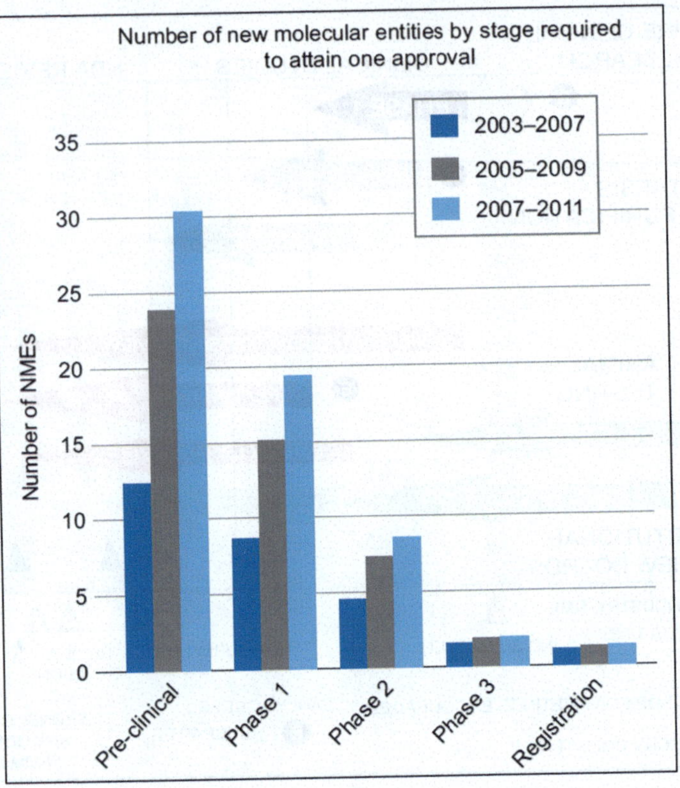

Fig. 3.19 Declining number of new drugs reaching approval. (Burrill, 2014) (Reprinted with permission from Elsevier)

slowed the process of bringing drugs through R&D to approval. In the 2000s, that time grew to 13.5 years from just 6 years in the 1970s (Burrill, 2014). A study by the consulting firm Bain & Company reported that the cost of discovering, developing, and launching (which factored in marketing and other business expenses) a new drug (along with the prospective drugs that fail) rose over 5 years to nearly $1.7 billion in 2003. According to Forbes, by 2010, development costs were between $4 billion to $11 billion per drug. Given the very high attrition rate, it is no surprise because it takes so much time, money, and resources to bring a new drug into the market successfully.

3.4.5 *Barriers to Commercial Launch*

National markets for pharmaceuticals differ on several dimensions, of which regulation is perhaps the most notable. The entry of pharmaceuticals is restricted, as is the price in many countries. Each nation has an agency or ministry devoted to pharmaceutical evaluation, which has heterogeneous standards for establishing safety and efficacy and varies in how quickly they evaluate new drug applications. Some require that some clinical trials be performed on domestic patients and are less accepting of foreign data. Some European countries require proof of cost-effectiveness. Price regulation has many variants such as control of manufacturer price, reimbursement practices, promotional spending control, wholesale and pharmacy margins, etc. (Kyle, 2007). Price controls refer to a cap on the ex-manufacturer price or the amount a national health service pays for a pharmaceutical product (the reimbursement price). The price for a drug is officially based on some determination of therapeutic value, the cost of comparable treatments, the contribution of the drug's manufacturer to the domestic economy, and manufacturing cost; the weight given to each factor differs by country. Notably, several countries reference the price of a drug in other countries when setting the price. This *international reference pricing* means that the price in one country can affect the price in other markets. Negotiations between pharmaceutical firms and national governments may be lengthy and tense, and drug companies often blame this process for delays in product launches (Kyle, 2007).

3.4.6 Patent Protections of Innovative Products

Two types of patent protection exist for innovative pharmaceutical companies. The first are traditional patents granted and processed by the United States Patent and Trademark Office (USPTO). When a drug is first discovered, the pharmaceutical company that discovers and markets it receives a patent on its new drug. The patent usually lasts 20 years from the application's filing date to allow the originating company to recoup its research investment. Traditional patents can be filed anywhere along the development lifeline of a drug, and they can encompass a wide range of claims. The second type of protection is regulatory and granted by the FDA upon approval. This type of protection is termed "exclusivity" and could run concurrently with traditional patent protection. For an NCE, exclusivity is granted 5 years after the FDA approval date. Orphan (for rare diseases and disorders) and pediatric drugs are granted more extended exclusivity periods because of their limited use. For example, orphan drugs are granted 7 years of exclusivity protection. Similarly, developing a product for the pediatric population provides the innovator's company with an additional 6 months of exclusivity.

Exclusivity became a critical provision in the Drug Price Competition and Patent Term Act of 1984—commonly called the Hatch–Waxman Act. The purpose of exclusivity was to provide pharmaceutical companies with 5 years of marketing protection, during which other manufacturers were prohibited from applying to sell a generic product. Exclusivity does not, however, prevent other manufacturers from seeking approval for a drug that uses the exact therapeutic mechanism as an already approved drug. These products are often called "me-too" drugs and must undergo the same rigorous clinical testing to garner FDA approval (Higgins & Rodriguez, 2006). The exclusivity provision was designed to help firms with no patent protection or little time left under a patent when the FDA approved an NCE. The loss of patent protection brings the most significant loss of earnings for an innovative pharmaceutical company and paves the way for generics.

3.5 Generic Pharmaceutical Industry's Perspective—Allow Competition Early

A generic drug is a pharmaceutical drug equivalent to a brand-name product in dosage, strength, route of administration, quality, performance, and intended use but does not carry the brand name. A generic drug must contain the same active ingredients as the original brand-name formulation. The FDA requires that generic drugs undergo a rigorous review process to receive approval so that a generic medication provides the same clinical benefit and is as safe and effective as the brand-name medicine that it duplicates. Generally, generic products become available after the patent protections afforded to a drug's original developer expire. Generic medicines tend to cost less than their brand-name counterparts because they do not have to repeat animal and clinical studies required of the brand-name medicines to demonstrate safety and effectiveness. When multiple generic companies market the same product, market competition typically results in prices about 85% lower than the brand name (FDA, 2018). As per the information provided by the Association for Accessible Medicines (AAM), generics account for 90% of prescriptions dispensed in the United States but only 17.5% of the total drug costs (AAM, 2023). However, this would not have been possible without a significant legislative act in the 1980s—The Hatch-Waxman Act.

3.5.1 ANDA and Hatch-Waxman Act

The Drug Price Competition and Patent Term Restoration Act (Public Law 98-417), informally known as the Hatch-Waxman Act,[11] is a 1984 US federal law that encourages the manufacture of generic drugs by the pharmaceutical industry and established the modern system of government generic drug regulation in the United States. Although the Federal Food, Drug, and Cosmetic Act made it possible for generic companies to get regulatory approval for drugs by filing an Abbreviated New Drug Application (ANDA), in the early 1980s, it became clear that very few generics were coming to market. Congress studied the issue and realized that under patent and regulatory law, it was easy for innovator companies to make it difficult

[11] Representative Henry Waxman of California and Senator Orrin Hatch of Utah sponsored the Act.

for generic companies to file ANDAs successfully. The regulatory pathway to get ANDAs approved was irregular and uncertain. In response, the Hatch-Waxman Act was negotiated and enacted. Hatch-Waxman amended the Federal Food, Drug, and Cosmetic Act. Section 505(j) of the Act, codified as 21 USC §355(j), outlines the process for pharmaceutical manufacturers to file an ANDA for approval of a generic drug.

The Act facilitates the filing of ANDAs by generic companies by preventing the FDA from asking a generic company to provide anything other than information on how it is going to manufacture the drug, quality assurance, and a study showing that the drug acts the same in a human as the innovator drug; this is called bioequivalence. This part of the Act is one of few pieces of legislation that restricts the powers and reach of a federal agency. The Act also gives generic companies safe harbor from patent infringement lawsuits during the time when the generic company is preparing its ANDA; during that time, the generic company needs to learn how to manufacture the drug, manufacture a test batch, and run bioequivalence studies, all activities for which it could be sued for infringement. This protection is called the research exemption.

3.5.2 *Race to First Generics*

To market a generic drug within the US, a company must file an ANDA with the FDA. Once approved, an applicant may manufacture and market the generic drug product to provide a safe, effective, low-cost alternative to the American public. An important measure of the ANDA process is the approval of the "first generics." First Generics are the first ones approved by the FDA, which permits a manufacturer to market a generic drug product in the United States. FDA considers first generics important to public health and prioritizes review of these submissions.

A key performance indicator for monitoring the pharmaceutical industry's productivity is to track the number of novel drugs and first generics approved within a given year. From 2013–2017, the number of novel drugs approved ranged from 22–46. In the same period, the number of first generics approved ranged from 73–106 (Mittal, 2019). This clearly shows that the pace of novel drug development is much slower than that of generics development, primarily due to the significant number of clinical trials that novel drug manufacturers must do to establish safety and efficacy. Another important business criterion to monitor is how quickly a novel drug could be made available into a first generic. It can take 6 to 30+ years for a novel drug to be available in the first generic. The reason for this large variance can be attributed to patent protections around a novel drug, difficulty establishing the safety and efficacy of a generic compared to the novel drug, or any other significant technical knowledge that keeps the generic competition at bay.

3.6 Patients' Perspective—Make Medicines Affordable with Proven Outcomes

Patients are at the heart of the healthcare debate. After all, they are the consumers of the products manufactured by the pharmaceutical industry and regulated by the government. There is no question that the government and the industry are passionate about providing the best care possible to patients. Growing research interest involves patients and other stakeholders as partners in study planning, conducting, and sharing information about health research results (Heckert et al., 2020). While the cost and affordability of medicines have always taken center stage to integrate patients' concerns in the healthcare debate, throughout the world, there is a growing recognition that the patient's perspective is highly relevant to efforts to deliver high-value patient-centered care and to improve the quality and effectiveness of healthcare.

One of the key challenges to achieving this is the limited measurement of outcomes that matter most to patients. The introduction of patient-reported outcome measures (PROMs)—measurement instruments designed to assess the status of a patient's health condition that comes directly from the patient—is one strategy to ensure that patient's perspectives are systematically incorporated into the approaches to delivering healthcare services and valuing the performance of the healthcare system. PROMs could improve the quality of care in the same way as any other benchmarking tool, and some suggest that PROMs can potentially transform healthcare (Dean et al., 2021). In this section, a detailed discussion of the patient's concerns regarding affordability, value, product design, and PROM is presented.

3.6.1 *Triple Aim Framework and PROM*

In the 1990s, reports in several countries revealed a staggering number of patient injuries and deaths each year due to avoidable errors and deficiencies in healthcare, among them adverse events and complications arising from poor infection control. In the United States, a 1999 report from the Institute of Medicine called for a broad national effort to prevent these events, including establishing patient safety centers, expanding reporting of adverse events, and developing safety programs in healthcare organizations. This led to the creation of multiple Patient Safety Organizations (PSOs), which can be defined as a group, institution, or association that improves medical care by reducing medical errors.

The standard functions of PSOs are data collection and analysis, reporting, education, funding, and advocacy. Although many PSOs are funded and run by governments, others have sprung from private entities such as industry, professional, and consumer groups. One such organization is the Institute for Healthcare Improvement (IHI), an independent not-for-profit organization helping to lead the improvement of health and healthcare throughout the world. Founded in 1991 and based in

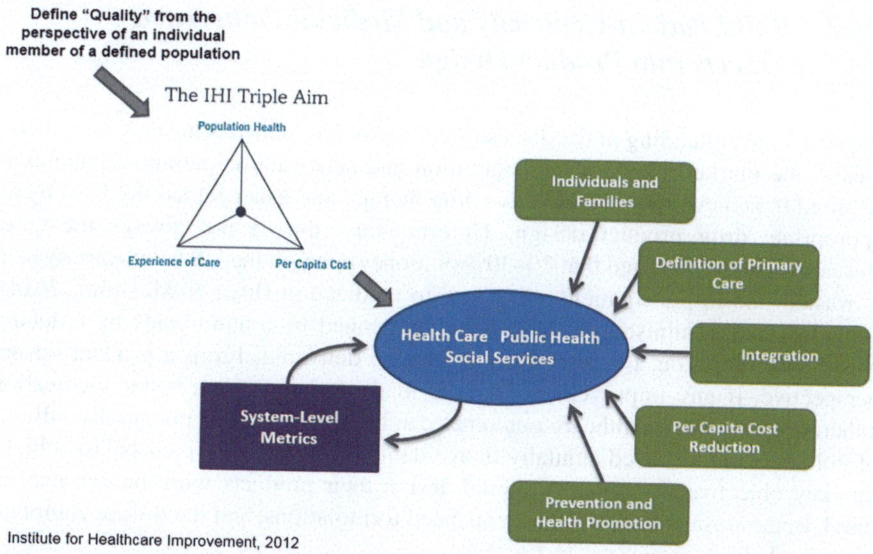

Fig. 3.20 Design of a triple aim enterprise. (IHI, 2022)

Boston, Massachusetts, IHI works to accelerate improvement by building the will for change, cultivating promising concepts for improving patient care, and helping healthcare systems implement those ideas. A vital heuristic of IHI is the Triple Aim (Fig. 3.20).

The IHI Triple Aim framework describes an approach to optimizing health system performance. IHI believes that new healthcare policies should be designed and developed to pursue three dimensions simultaneously: (1) improving the patient experience of care (including quality and satisfaction), (2) improving the health of populations, and (3) reducing the per capita cost of healthcare (Berwick et al., 2008). In most healthcare settings today, no one is accountable for all three dimensions of the IHI Triple Aim. Because the Triple Aim entails ambitious improvement at all levels of the system, IHI advocates a systematic approach to change that includes the identification of target populations, definition of system aims and measures, development of a portfolio of project work that is sufficiently strong to move system-level results, and rapid testing and scale up that is adapted to local needs and conditions. In the US environment, many areas of health reform can be furthered and strengthened by Triple Aim thinking, including accountable care organizations (ACOs), bundled payments, and other innovative financing approaches; new models of primary care, such as patient-centered medical homes; sanctions for avoidable events, such as hospital readmissions or infections; and the integration of information technology (IHI, 2022).

3.6.2 *Build Patient-Centricity and Medicine Adherence Aspects into Product Design*

Improved understanding of the disease mechanism has helped bring new drug therapies to the market. However, collaboration and active participation of patients is required to achieve safe and effective drug therapy and use, enabled not least by an appropriate drug product design. Unfortunately, this is not always the case. Researchers have asserted that 20–30% of money spent in the US healthcare system is wasted due to poor patient adherence to medication (Iuga & McGuire, 2014). Providers and administrators have been challenged to contain costs by reducing waste and improving the effectiveness of care delivered. From a product design perspective, if any improvements can be made to allow for a better medication adherence scenario, healthcare outcomes can be improved, and potentially billions of dollars could be saved annually in avoidable direct healthcare costs. To achieve this key objective, formulators should design their products with the end-user in mind. Better dosage form designs, enhanced formulations, and fixed-dose combinations can help.

Patients are considered adherent to medications when they take prescribed agents at doses and times recommended by a healthcare provider and agreed to by the patient. Medication nonadherence is a significant public health consideration, leading to poor health outcomes, increased healthcare service utilization, and healthcare costs. Research has shown that poor medication adherence is widespread and varied by disease, patient characteristics, and insurance coverage. Studies have consistently shown that 20% to 30% of medication prescriptions are never filled and that approximately 50% of medications for chronic diseases are not taken as prescribed. This lack of adherence has dramatic effects on health. In the United States, it is estimated to cause approximately 125,000 deaths, at least 10% of hospitalizations, and a substantial increase in morbidity and mortality per year. Non-adherence has been estimated to cost the US healthcare system between $100 billion and $289 billion annually (Viswanathan et al., 2012).

The problem of poor patient adherence to medication is, unfortunately, a global problem. In a report by WHO from 2003, it was reported that adherence among patients suffering from chronic diseases averages only 50% in developed countries. The magnitude and impact of poor adherence in developing countries are assumed to be even higher, given the paucity of health resources and inequities in access to healthcare. For example, in China, Gambia, and Seychelles, only 43%, 27%, and 26%, respectively, of patients with hypertension adhere to their antihypertensive medication regimen. Data on patients with depression reveal that between 40% and 70% adhere to antidepressant therapies. In Australia, only 43% of the patients with asthma take their medication as prescribed all the time, and only 28% use prescribed preventive medication. In the treatment of HIV and AIDS, adherence to antiretroviral agents varies between 37% and 83%, depending on the drug under study and the demographic characteristics of patient populations. This represents a tremendous challenge to population health efforts where success is determined primarily by adherence to long-term therapies (WHO, 2010). It is no wonder that providers and administrators have been challenged to contain costs by reducing waste and improving the effectiveness of care delivered. Patient nonadherence to prescribed medications is associated with poor therapeutic outcomes, progression of the disease, and a significant financial burden due to avoidable direct healthcare costs.

From the pharmaceutical dosage form design viewpoint, an outcome-orientated definition is more applicable. There are numerous reasons for non-compliance from a patient when dealing with an outcome-oriented approach for a solid oral dosage form, ranging from the length of therapy, dosage form size, product packaging, and the physical similarity of different products. No matter the reason for a patient's poor (or complete lack of) compliance, the costs incurred are huge! The study of patient adherence to medication is a subject of healthcare research that involves studying of patient-related factors (such as demographics, sociocultural, and behavioral) and external factors (such as disease, medication, provider communication, and access to care). From a patient's perspective, there are myriad reasons for nonadherence (or poor adherence) to medication. However, pharmaceutical manufacturing can aid in improving healthcare outcomes by incorporating certain aspects of patient-centricity in their products' design (Stegemann et al., 2016; Mittal, 2019).

Patient centricity can be defined as integrated measures for listening to and partnering with patients and placing patient well-being at the core of all initiatives. In essence, it represents a holistic approach to disease management. This is a relatively new concept today, and the patients are more knowledgeable than ever. Recent technological and scientific advances have enabled easy access to information; patients are online and can connect with peers, exchange experiences and knowledge in real time, participate in advocacy and patient support groups, and contribute to real-time data generation. As a result, patients feel rightly empowered and expect their voices to be heard. Although such a discussion has typically taken place in the context of clinical outcomes, pharmaceutical manufacturing is embracing the concept of patient-centricity in product design. With real-life patient usage issues such as multi-morbidity and the ensuing polypharmacy, incorporating patient-centric dosage form design concepts is a value-added proposition that can be built into the product early on. Therefore, as part of developing commercial products, product design teams must also conduct a comprehensive assessment of patient characteristics such as disease conditions, disabilities, impairments, and specific needs. From a business perspective, incorporating patient-centric design features allows them to differentiate themselves from the competition and help make their products more desirable in the marketplace. Undoubtedly, innovation plays a grand role in achieving this desired state of success.

3.6.3 *Develop New Products for Special Patient Populations*

Medicinal science has made tremendous progress over the last 20 years. These advances have supported the introduction of new drug therapies to improve therapeutic outcomes and tackle life-threatening diseases for which no treatment previously existed. The overall increase in life expectancy due to advances in pharmacotherapy, hygienic measures, healthcare, and wealth contributes to an increase in the number and type of special patient populations (SPP). Examples of SPP include pregnant women, pediatrics, critically ill, or geriatrics. The SPPs differ from the traditional adult population, e.g., concerning their clinical presentation and physical, physiological, or psychological patterns. This creates some limitations to developing drugs for SPP.

Patients in clinical trials during drug development are selected based on strict inclusion and exclusion criteria. Consequently, efficacy and safety are understudied and poorly characterized in SPP for most drugs on the market. As a result, dosing regimens are not selected for optimal efficacy or minimal toxicity in these patients. Also, SPP is often smaller than reference populations. Especially in children, practical and ethical constraints also limit the number of patients that can be included in studies and the number of invasive samples that can be obtained per patient.

Moreover, studies in special populations are generally performed during routine clinical practice, which may negatively impact the information content of the obtained data. As a result, data obtained in special patient populations requires advanced methods with increased flexibility and statistical power to obtain a sufficiently high level of evidence from the pharmacokinetic and pharmacodynamics data to derive drug dosing regimens (Krekels et al., 2017). Most of the products for special patient populations require reformulation or adaptation of new drug delivery technologies.

3.7 Summary—Moderating the Debate

Access to medicines is a human right. Unfortunately, in a debate around access to medicines, an easy target is the high cost of pharmaceuticals. For example, Americans pay higher healthcare prices than people in other developed countries. That is true of prices for doctors, especially proceduralists and specialists, and non-physician goods and services such as drugs, imaging, and technologies. Also, there is variation in drug prices within the United States: some Americans pay much more than others for the same product. These appalling disparities are generally not related to quality, outcomes, or patient population but are instead driven by the provider's market power (Khullar & Rajkumar, 2018). However, pharmaceutical drug pricing is only one aspect of this discussion. Another aspect is to address the inefficiencies and wastage in the healthcare systems. In the same vein, it is equally important to understand how pharmaceutical products are developed, regulated, and the challenges that they face in commercial markets. It is, therefore, critical to review the steps that the pharmaceutical industry has taken (and continues to take) to make its products more affordable without compromising quality. ***Such an evaluation can help in enriching the healthcare debate with facts and can hope to find potential solutions regarding high pharmaceutical pricing. Couldn't it?***

References

AAM. (2023). *Generic medicines*. Retrieved March 4, 2024 from https://www.accessiblemeds.org/generic-medicines

AHP. (2022). *US pharmaceutical pricing: An overview*. Axene Health Partners, LLC. Retrieved September 9, 2022 from https://axenehp.com/us-pharmaceutical-pricing-overview/

Arrow, K. J. (1963). Uncertainty and the welfare economics of medical care. *The American Economic Review, 53*(5), 941–973. http://www.jstor.org/stable/1812044

BEA. (2022). *Consumer spending*. BEA. Retrieved August 20, 2022 from https://www.bea.gov/resources/learning-center/what-to-know-consumer-spending

Berwick, D. M., Nolan, T. W., & Whittington, J. (2008). The triple aim: Care, health, and cost. *Health Affairs, 27*(3), 759–769. https://doi.org/10.1377/hlthaff.27.3.759

Burrill, G. S. (2014). Chapter 3—The biotechnology industry: An engine of innovation. In C. Shimasaki (Ed.), *Biotechnology entrepreneurship* (pp. 21–44). Academic. https://doi.org/10.1016/B978-0-12-404730-3.00003-8

Chokshi, M., Patil, B., Khanna, R., Neogi, S. B., Sharma, J., Paul, V. K., & Zodpey, S. (2016). Health systems in India. *Journal of Perinatology, 36*(3), S9–S12. https://doi.org/10.1038/jp.2016.184

CMS. (2019). *CMS fast facts*. Retrieved from https://www.cms.gov/Research-Statistics-Data-and-Systems/Statistics-Trends-and-Reports/CMS-Fast-Facts/index

CMS. (2021). *Nation's health dollar: Where it came from, where it went*. Retrieved from https://www.cms.gov/Research-Statistics-Data-and-Systems/Statistics-Trends-and-Reports/NationalHealthExpendData/NationalHealthAccountsHistorical

Danzon, P. M. (2014). Pricing and reimbursement of biopharmaceuticals and medical devices in the USA. In A. J. Culyer (Ed.), *Encyclopedia of health economics* (pp. 127–135). Elsevier. https://doi.org/10.1016/B978-0-12-375678-7.01209-8

De Pietro, C., Camenzind, P., Sturny, I., Crivelli, L., Edwards-Garavoglia, S., Spranger, A., Wittenbecher, F., & Quentin, W. (2015). Switzerland: Health system review. *Health Systems in Transition, 17*(4), 1–288. xix.

Dean, S., Al Sayah, F., & Johnson, J. A. (2021). Measuring value in healthcare from a patients' perspective. *Journal of Patient-Reported Outcomes, 5*(2), 88. https://doi.org/10.1186/s41687-021-00364-4

Desai, M. (2006). Strong medicine: The pharmaceutical Industry's compact with society. *Georgetown Journal of International Affairs, 7*(1), 93–99. http://www.jstor.org/stable/43133665

Dey, P., & Bach, P. B. (2019). The 6 functions of health insurance. *JAMA, 321*(13), 1242–1243. https://doi.org/10.1001/jama.2019.2320

Dubois, P., De Mouzon, O., Scott-Morton, F., & Seabright, P. (2015). Market size and pharmaceutical innovation. *The Rand Journal of Economics, 46*(4), 844–871.

FDA. (1998). *The CDER handbook*.

FDA. (2018). *Generic drug facts*. FDA. https://www.fda.gov

Fuchs, V. R. (1972). *Essays in the economics of health and medical care*.

Gay, E. F. (1932). The great depression. *Foreign Affairs, 10*(4), 529–540. https://doi.org/10.2307/20030459

Grossman, M. (1972). *The demand for health: A theoretical and empirical investigation*. National Bureau of Economic Research (NBER).
Halpenny, G. M. (2016). High drug prices hurt everyone. *ACS Medicinal Chemistry Letters, 7*(6), 544–546. https://doi.org/10.1021/acsmedchemlett.6b00139
Heckert, A., Forsythe, L. P., Carman, K. L., Frank, L., Hemphill, R., Elstad, E. A., Esmail, L., & Lesch, J. K. (2020). Researchers, patients, and other stakeholders' perspectives on challenges to and strategies for engagement. *Research Involvement and Engagement, 6*(1), 60. https://doi.org/10.1186/s40900-020-00227-0
Higgins, M. J., & Rodriguez, D. (2006). The outsourcing of R&D through acquisitions in the pharmaceutical industry. *Journal of Financial Economics, 80*(2), 351–383. https://doi.org/10.1016/j.jfineco.2005.04.004
IHI. (2022). *The IHI triple aim*. Institute for Healthcare Improvement. Retrieved September 10, 2022 from https://www.ihi.org/Engage/Initiatives/TripleAim/Pages/default.aspx
IQWiG. (2018). *Health care in Germany: The German health care system*. https://www.ncbi.nlm.nih.gov/books/NBK298834/
Iuga, A. O., & McGuire, M. J. (2014). Adherence and health care costs. *Risk Management and Healthcare Policy, 7*, 35–44. https://doi.org/10.2147/RMHP.S19801
Jenner, B. C. (2010). *Health insurance industry market structure: Coverage and cost issues*. Nova Science Publishers, Incorporated. http://ebookcentral.proquest.com/lib/claremont/detail.action?docID=3019736
Khullar, D., & Rajkumar, R. (2018). Innovations to cut US healthcare prices. *BMJ: British Medical Journal, 361*. https://www.jstor.org/stable/26960650
Kleinrock, M., & Muñoz, E. (2020). *IQVIA report on global medicine spending and usage trends*. https://www.iqvia.com/-/media/iqvia/pdfs/institute-reports/global-medicine-spending-and-usage-trends.pdf?_=1660519466448
Koba, M. (2011). *Medicare and medicaid: CNBC explains*. https://www.cnbc.com/id/43992654
Krekels, E. H. J., van Hasselt, J. G. C., van den Anker, J. N., Allegaert, K., Tibboel, D., & Knibbe, C. A. J. (2017). Evidence-based drug treatment for special patient populations through model-based approaches. *European Journal of Pharmaceutical Sciences, 109s*, S22–S26. https://doi.org/10.1016/j.ejps.2017.05.022
Kronick, R. (2016). AHRQ's role in improving quality, safety, and health system performance. *Public Health Reports, 131*(2), 229–232. https://doi.org/10.1177/003335491613100205
Kyle, M. K. (2007). Pharmaceutical price controls and entry strategies. *The Review of Economics and Statistics, 89*(1), 88–99. http://www.jstor.org/stable/40043076
Mazie, D. M. (2024). Medicare and Medicaid. In *Encyclopedia Britannica*.
Mittal, B. (2019). How to integrate quality by efficient design (QbED). In *Product development*. Elsevier Academic Press.
OECD. (2021). *Health at a glance 2021: OECD indicators*. OECD Publishing. https://doi.org/10.1787/ae3016b9-en
Papanicolas, I., Woskie, L. R., & Jha, A. K. (2018). Health care spending in the United States and other high-income countries. *JAMA, 319*(10), 1024–1039. https://doi.org/10.1001/jama.2018.1150
Park, J. H., & Shin, K. (2018). R&D project valuation considering changes of economic environment: A case of a pharmaceutical R&D project. *Sustainability, 10*(4), 993. https://www.mdpi.com/2071-1050/10/4/993
Psotka, M. A., Fiuzat, M., Solomon, S. D., Chauhan, C., Felker, G. M., Butler, J., Teerlink, J. R., Sinha, S. S., O'Connor, C. M., & Konstam, M. A. (2020). Challenges and potential improvements to patient access to pharmaceuticals. *Circulation, 142*(8), 790–798. https://doi.org/10.1161/CIRCULATIONAHA.119.044976
Rice, T., Rosenau, P., Unruh, L. Y., Barnes, A. J., Saltman, R. B., & van Ginneken, E. (2013). United States of America: Health system review. *Health Systems in Transition, 15*(3), 1–431.
Samanen, J. (2013). Chapter 7 - The structure and business of biopharmaceutical companies including the management of risks and resources. In R. Ganellin, S. Roberts, & R. Jefferis

(Eds.), *Introduction to Biological and Small Molecule Drug Research and Development* (pp. 225–253). Elsevier. https://doi.org/10.1016/B978-0-12-397176-0.00007-8
Shrank, W. H., DeParle, N.-A., Gottlieb, S., Jain, S. H., Orszag, P., Powers, B. W., & Wilensky, G. R. (2021). Health costs and financing: Challenges and strategies for a new administration. *Health Affairs, 40*(2), 235–242. https://doi.org/10.1377/hlthaff.2020.01560
Silverman, R. B. (2004). *The organic chemistry of drug design and drug action*. Academic.
Stegemann, S., Ternik, R. L., Onder, G., Khan, M., & van Riet-Nales, D. (2016). Defining patient centric pharmaceutical drug product design. *The AAPS Journal, 18*, 1047–1055. https://doi.org/10.1208/s12248-016-9938-6
Tang, N., Eisenberg, J. M., & Meyer, G. S. (2004). The roles of government in improving health care quality and safety. *Joint Commission Journal on Quality and Safety, 30*(1), 47–55. https://doi.org/10.1016/s1549-3741(04)30006-7
Thomas, L. (2021). *Healthcare systems around the world*. https://www.news-medical.net/health/Healthcare-Systems-Around-the-World.aspx
Various. (2022). *Health care systems by country*. Wikipedia. Retrieved September 7, 2022 from https://en.wikipedia.org/wiki/Health_care_systems_by_country#cite_note-3
Varol, N., Costa-Font, J., & McGuire, A. (2012). Does adoption of pharmaceutical innovation respond to changes in the regulatory environment? *Applied Economic Perspectives and Policy, 34*(3), 531–553.
Vincent Rajkumar, S. (2020). The high cost of prescription drugs: Causes and solutions. *Blood Cancer Journal, 10*(6), 71. https://doi.org/10.1038/s41408-020-0338-x
Viswanathan, M., Golin, C. E., Jones, C. D., et al. (2012). Interventions to improve adherence to self-administered medications for chronic diseases in the united states: A systematic review. *Annals of Internal Medicine, 157*(11), 785–795. https://doi.org/10.7326/0003-4819-157-11-201212040-00538
WHO. (2010). *The world health report: Health systems financing—The path to universal coverage*. Retrieved from https://www.who.int/publications/i/item/9789241564021
World101. (2023). *Global health: How health care works around the World*. Retrieved March 3, 2024 from https://world101.cfr.org/global-era-issues/global-health/how-health-care-works-around-world#:~:text=Around%20the%20world%2C%20health%2Dcare,can%20access%20and%20afford%20them

Chapter 4
Drug Pricing and Reimbursement

Competition is not only the basis of protection to the consumer, but is the incentive to progress

Herbert Hoover

Contents

Abstract The pharmaceutical industry has a profitable but precarious operating model. Until the product launch, the cash flow is negative due to the R&D expenses. Upon launch, the drug sales bring in revenues, and cash flow becomes positive. This business model has been around for a long time, but in recent times, the prices and profits in the biopharmaceutical industry have been increasingly scrutinized in social debate. With the passage of the Inflation Reduction Act of 2022 and a concerted effort worldwide to move towards value-based pricing systems, the marketplace for pharmaceuticals is evolving rapidly as most governments struggle to balance health and social expenditures at sustainable levels. Therefore, the product designers must be aware of the market dynamics that might most likely exist when their products enter the marketplace. The practical reality is that the pharmaceutical industry may need to account for a future where value-based pricing is the norm and

B. Mittal, *Sustainable Global Health Systems and Pharmaceutical Development*, AAPS Advances in the Pharmaceutical Sciences Series 60,
https://doi.org/10.1007/978-3-031-50415-0_4

not the exception. In such a scenario, designers of pharmaceutical products will routinely be asked what can be (further) done to reduce the costs so that the product does not cross the cost-effectiveness threshold. Situational awareness of the ever-changing and evolving healthcare marketplace is thus paramount for pharmaceutical scientists.

Keywords Economic value · Medicare · Medicaid · Inflation reduction act · Medicare price negotiation · Pharmacoeconomics · Cost-effectiveness threshold · Incremental cost-effectiveness ratio · Value-based pricing · Quality-adjusted life year

4.1 Introduction

Economics studies how people use limited resources (such as time, money, and labor) to produce, exchange, and consume goods and services. It explains how these scarce resources are allocated among competing ends by the economic system. Economists typically use the concept of the production possibilities frontier (PPF) to illustrate the function of scarcity, choice, and opportunity costs (Ruffin & Gregory, 2001). A typical PPF is shown in Fig. 4.1, where a firm (or a country) may have to evaluate information about its opportunity cost when producing more than one good or service. Knowing the precise opportunity cost at each junction helps managers achieve production efficiency (Wolff, 2022).

For example, in Fig. 4.1, any spot inside the curve (point D) reflects inefficient production since not all the resources are utilized. On the other hand, any area beyond

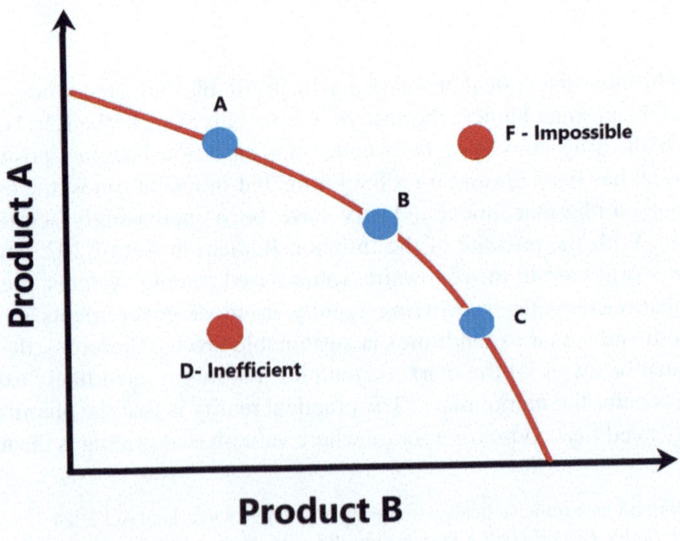

Fig. 4.1 Evaluation of PPFs

the curve (point F) is impossible to achieve because of its resource constraints. In contrast, all along the actual curve (A, B, C) is the possibility frontier of production, where each point demonstrates the opportunity cost as every additional unit of one good decreases the output of the other good. In Fig. 4.1, an outward shift will illustrate greater possible output. For instance, total employment would increase all possible combinations of production. Similarly, increased skilled labor, new technology, or access to more capital goods will shift the curve outwards. An inward shift would illustrate a decrease in possible production. Many factors can shift a country's PPC inward. For instance, high unemployment, rising inflation, and outflow of capital goods will all cause a shift inwards, illustrating less possible output (Wolff, 2022).

While the PPF provides a visualization of the resource allocation and associated efficiencies of production, an important coordinating mechanism is that of the price system. The price system provides necessary information for informed decision-making, allowing resource owners to trade freely, buying and selling at whatever relative price emerges in the marketplace. The price systems lead to the concept of economic value. Economic value is defined as how well the service or product serves its intended purpose at a price the customers are willing to pay. Any service provider (or manufacturer) has to balance its competitive priorities of top quality vs. low-cost operations to produce value for the customer. How much economic value a service or product has in the customer's mind depends on their expectations before purchasing it (Krajewski et al., 2010). One possible outcome of an out-of-control valuation of any asset is that it can lead to an economic bubble.

An economic bubble (sometimes referred to as a "speculative bubble" or "market bubble") is trading in high volumes at prices that are considerably at variance from intrinsic values. The intrinsic value is a theoretical calculation that aims to reflect the fair value by considering hypotheses of future returns and risks. The cause of bubbles remains a challenge to economic theory. The main idea behind creating economic bubbles is a weak financial policy and excessive monetary liquidity in the financial system (Girdzijauskas et al., 2009). There have been a few economic bubbles in the last few decades: the 1980s Japan Real Estate and Stock Market bubble, the 1990s Dotcom bubble, and the early 2000s US housing bubble. However, one classic example of an economic bubble is the Dutch tulip mania in 1634–37 markets.

4.1.1 Philosophical Digression—A Blast from the Past

There is a cautionary tale often recited to young investors who wish to take advantage of the volatility of certain portions of the stock market in ways that seem untraditional for seasoned investors. In speculative markets, prices of certain stocks fluctuate rapidly. Securities, or stocks and bonds, have price points changing every minute, if not seconds, and speculators attempt to gain an advantage during this volatility. Investments in such speculative markets are not new. Almost 400 years ago, the Dutch tulip mania was the first recorded speculative bubble or asset bubble in history. The tulip mania was a period during the Dutch Golden Age when contract prices for some bulbs of the recently introduced and fashionable tulip reached extraordinarily high levels. The major acceleration started in 1634 and then dramatically collapsed in February 1637.

The tulip (genus *Tulipa*), genus of about 100 species of bulbous herbs in the lily family (Liliaceae), is native to Central Asia and Turkey. Tulips are among the most popular garden flowers, and numerous cultivars and varieties have been developed over the centuries. Tulip cultivation likely began in Persia (Iran) in the tenth century, eventually becoming a symbol of the Ottoman Empire. Tulips were introduced to the Western world in the mid-1500s, and by 1562, a nascent horticultural tulip industry had started in Europe. The tulip produces two or three thick bluish-green leaves clustered at the plant's base. The usually solitary bell-shaped flowers have three petals and three sepals. There are six free stamens, and the three-lobed ovary is terminated by a sessile three-lobed stigma. The fruit is a capsule with many seeds. Many garden tulips can be propagated only by their scaly bulbs (Britannica, 2024a).

The demand for differently colored varieties of tulips soon exceeded the supply, and prices for individual bulbs of rare types began to rise to unwarranted heights in northern Europe. Over the years, researchers have delved deeper into the origins of this medieval phenomenon. In one such article, researchers posit that it is important to consider tulip bulbs as investment goods to understand this speculative market. The tulip flower has extraordinarily vivid colors and variations that appear in strikingly distinct flames and flares; these explain the demand for the flowers. Nevertheless, wholesale trading was always in bulbs, and the blossoming flower was never the object of wholesale trade. Therefore, it was neither the prices of tulip flowers nor the prices of tulip seeds, but the bulb prices that were the subject of speculations leading to dramatic price increases in early fall 1636 and the subsequent upward-accelerating prices on claims for future bulbs that inspired the term "tulip mania". Tulip bulbs, rather than tulip flowers, were the object of wholesale trade because bulbs, unlike seeds or flowers, are an economically viable investment good. Bulbs produce annual underground offshoots, or offsets that grow into new bulbs, which grow more offsets, become new bulbs, etc. Existing bulbs, those that have flowered in the past, if planted in October/November, will again produce two to three offset bulbs that are accessible when harvested the following summer. Although seeds can produce bulbs that will eventually flower, growing tulips from seeds is not a good substitute (neither economically nor horticulturally) compared to growing from existing bulbs (McClure & Thomas, 2017).

The tulip bulb craze reached its height in Holland during 1633–37. Before 1633, Holland's tulip trade had been restricted to professional growers and experts, but the steadily rising prices tempted many ordinary middle-class and low-income families to speculate in the tulip market. Homes, estates, and industries were mortgaged so that bulbs could be bought for resale at higher prices. Sales and resales were made many times over without the bulbs ever leaving the ground, and rare varieties of bulbs sold for the equivalent of hundreds of dollars each. The crash came early in 1637 when doubts arose about whether prices would continue to increase. Almost overnight, the price structure for tulips collapsed (Fig. 4.2), sweeping away fortunes and leaving behind financial ruin for many ordinary Dutch families (Britannica, 2024b).

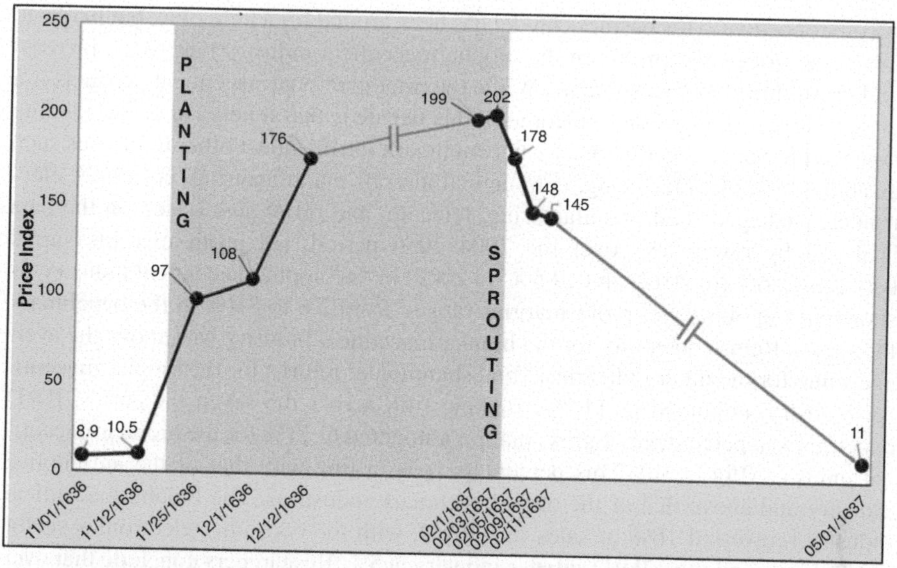

Fig. 4.2 Prices of Dutch tulip bulbs from 1636 to 1637. (McClure & Thomas, 2017) (Reprinted with permission from Cambridge University Press)

4.1.2 *Opportunity for Pharmaceutical Scientists: Anticipate the Dynamics of the Healthcare Marketplace*

United States' global pharmaceutical leadership is built upon an integrated and collaborative system of academics, medical institutions, government research labs, and private enterprises that perform and support advanced R&D, resulting in innovative new treatments and cures. Organizations that deal with developing innovative products spend significant resources on drug discovery and development. Hundreds and sometimes thousands of chemical compounds must be made and tested to find one that can achieve a desirable result. Although estimates vary by indication, the FDA assesses that it takes approximately eight-and-a-half years to study and test a new drug before it can be approved for the public. This estimate includes early laboratory and animal testing and later clinical trials using human subjects. During this entire time, millions of dollars are expended to develop the drug, along with thousands of experiments done by various disciplines to prove the dosage form's safety, efficacy, manufacturability, and quality. This creates a scenario where the current state of development has to take into account a high rate of attrition during drug development. This, however, is only one part of the debate.

As discussed in Sect. 3.4.1, the pharmaceutical industry has a profitable but precarious operating model. Until the product launch, the cash flow is negative due to the R&D expenses. Upon launch, the drug sales bring in revenues, and cash flow becomes positive. This business model has been around for a long time, but in recent times, the prices and profits in the biopharmaceutical industry have been increasingly scrutinized in social debate. While the pricing of pharmaceutical products is a divisive topic, an important outcome of this debate is that it helps in understanding how the pharmaceutical industry can benchmark itself against other industries such as automotive manufacturing, commercial aircraft manufacturing, consumer electronics, packaged food manufacturing, telecom, and oil & gas. Based on the data analyzed by researchers over the 2004–2016 period, net profit margins varied between 12.6% (in 2014) and 19.5% (in 2009) in the biopharmaceutical industry, as shown in Fig. 4.3a. Net profit margins ranged from 2.6 to 8.4% in the benchmark industries. Return on equity for the biopharmaceutical industry was above the average value for the other industries. Total shareholder returns for the biopharmaceutical industry amounted to 11.7%, ranking fifth across the seven industries. R&D expenses as a percentage of gross margin amounted to 21% for the biopharmaceutical industry (Fig. 4.3b). This percentage was in line with that of the automotive industry and above that of the other benchmark industries. The biopharmaceutical industry reinvested 16% of sales into R&D, with the consumer electronics sector being the second most R&D-intense industry at 5%. Researchers conclude that over the 2004–2016 period, the biopharmaceutical industry did not attain a risk-adjusted return on investment above that of other industries and, thus, did not outperform these industries (Popa et al., 2018).

All of this data analysis helps us understand the economics of the pharmaceutical industry. However, the marketplace for pharmaceuticals is evolving rapidly as most governments struggle to balance health and social expenditures at sustainable levels. Therefore, the product designers must be aware of the market dynamics that might most likely exist when their products enter the marketplace. An understanding of the ever-changing and evolving healthcare marketplace is thus paramount for pharmaceutical scientists.

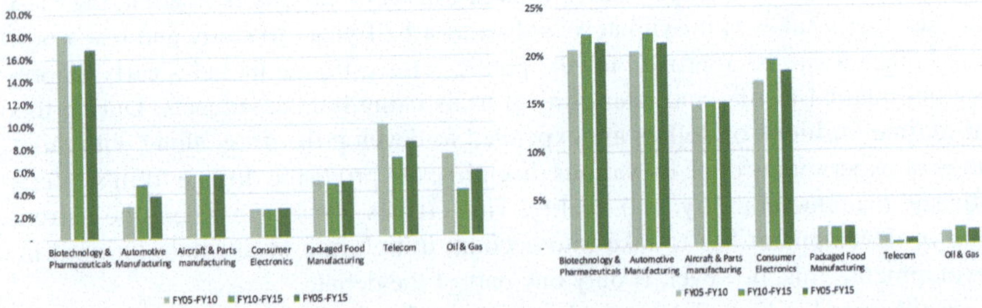

(a) Net Profit Margin **(b) R&D Expenses as Percentage of Gross Margin**

Fig. 4.3 Benchmarking of selected industries from 2004–2016 (Popa et al., 2018). (**a**) Net profit margin. (**b**) R&D expenses as percentage of gross margin

4.2 Traditional Models of Pharmaceutical Marketplace

The United States is the largest market for pharmaceuticals in the world. It is also one of the few countries in the industrialized world that does not regulate pharmaceutical prices. This primarily reflects the predominance of competing private health insurance plans in the US. Although the public insurance programs, mainly Medicare and Medicaid, now account for more than 40% of total health expenditures in the US, their design has historically been influenced by the design of the private insurance sector. In particular, Medicare Part D, which covers outpatient drugs for seniors, was designed to be run by competing private-sector prescription drug plans (PDPs), and the enabling legislation explicitly bars the government from negotiating drug prices (Danzon, 2014). It is important to understand the historical context here.

4.2.1 *Establishment of Medicare/Medicaid and Subsequent Reforms*

Most health insurance plans in the US are managed care plans (HMOs, PPOs) rather than indemnity or traditional health insurance plans that provide unlimited reimbursement for a fixed premium. Other parties involved in the health insurance market include employers (most private health insurance is obtained through an employer), federal, state, and local governments, and healthcare providers. At the federal level, the Centers for Medicare and Medicaid Services (CMS) administers both Medicare and Medicaid, the fundamental pillars of the US government's healthcare policy. President Lyndon B. Johnson signed the Social Security Amendments (Titles XVIII and XIX) on July 30, 1965, establishing Medicare and Medicaid, respectively. These amendments were made to the Social Security Act of 1935 and were enacted in 1966. A detailed timeline of key events that shaped the pharmaceutical regulations in the US is presented in Table 4.1. The latest statistics on the beneficiaries of Medicare services are provided in Fig. 4.4.

Why was prescription drug benefit not included in the original amendments of 1965?

The answer to the above question is hard to understand, even with 20/20 hindsight. As discussed in Table 3.2, Medicare has the following parts: Part A (Hospital Insurance), Part B (Medical Insurance), and Part D (Prescription Drug Coverage). Medicare Part D is insurance that helps lower prescription drug costs and protects against higher prices in the future. Part D was enacted as part of the Medicare Modernization Act of 2003 and went into effect on January 1, 2006. There have been efforts to add prescription drug coverage to the Medicare program earlier than 2003. In an excellent article, researchers have identified several vital patterns in policymaking over the past 60 years. They conclude three important reasons why

Table 4.1 Timeline of major US healthcare-related laws

Year	Event	Key provisions and impact
1906	Pure Food and Drug Act, signed into law by President Theodore Roosevelt	The Act created the FDA The Act prohibited, under penalty of seizure of goods, the interstate transport of food which had been "adulterated," with that term referring to the addition of fillers of reduced "quality or strength," coloring to conceal "damage or inferiority," formulation with additives "injurious to health," or the use of "filthy, decomposed, or putrid" substances
1938	Food, Drug, and Cosmetic Act (FD&C Act) signed into law by President Franklin Delano Roosevelt	The new law significantly increased federal regulatory authority over drugs by mandating a pre-market review of the safety of all new drugs, as well as banning false therapeutic claims in drug labeling without requiring that the FDA prove fraudulent intent The law also authorized factory inspections, expanded enforcement powers, set new regulatory standards for foods, and brought cosmetics and therapeutic devices under federal regulatory authority
1962	Kefauver-Harris Amendments to the 1938 FD&C Act	Sponsors of NDAs needed to document evidence of both safety and efficacy Informed consent was required of patients participating in clinical trials Sponsor should file an IND application with the FDA before initiating human testing Drug advertising should be required to disclose accurate information about side effects Mandated that a Drug Efficacy Study Implementation (DESI) be undertaken to classify all pre-1962 drugs that already were on the market as either effective, ineffective, or needing further study Clarified distinctions between brand and generic drugs and regulating their marketing
1965	Amendments to the Social Security Act of 1935	Establishment of Medicare and Medicaid At that time, Medicare covered only prescription drugs taken by hospital inpatients under Part A and physician-administered drugs (typically injections) under Part B
1984	Hatch-Waxman Drug Price Competition and Patent Term Restoration Act	It established the ANDA pathway that required only establishing bioequivalence with the reference drug (eliminating the need to establish safety and efficacy anew) and compliance with good manufacturing practices
2003	Medicare Modernization Act signed into law by President George W. Bush	Medicare part D was enacted as part of the act and went into effect on January 1, 2006

(continued)

Table 4.1 (continued)

Year	Event	Key provisions and impact
2010	Patient Protection and Affordable Care Act signed into law by President Barack Obama	It represents the US healthcare system's most significant regulatory overhaul and expansion of coverage since the enactment of Medicare and Medicaid in 1965 It amended the Public Health Service Act to create an abbreviated approval pathway for biological products that are demonstrated to be "highly similar" (biosimilar) to or "interchangeable" with an FDA-approved biological product. These new statutory provisions also may be referred to as the Biologics Price Competition and Innovation Act of 2009 (BPCI Act)
2022	Inflation Reduction Act signed into law by President Joe Biden	The IRA enacted the most significant reform for the payment of drugs and biologicals The IRA's Medicare-related provisions fall into two general categories: (1) reduction of prescription drug prices and (2) reduction of beneficiary cost sharing and premiums It is expected that over the 10 years between 2022 and 2031, through prescription drug negotiation and the Medicare inflationary rebates, approximately $101 billion and $71 billion, respectively, can be saved

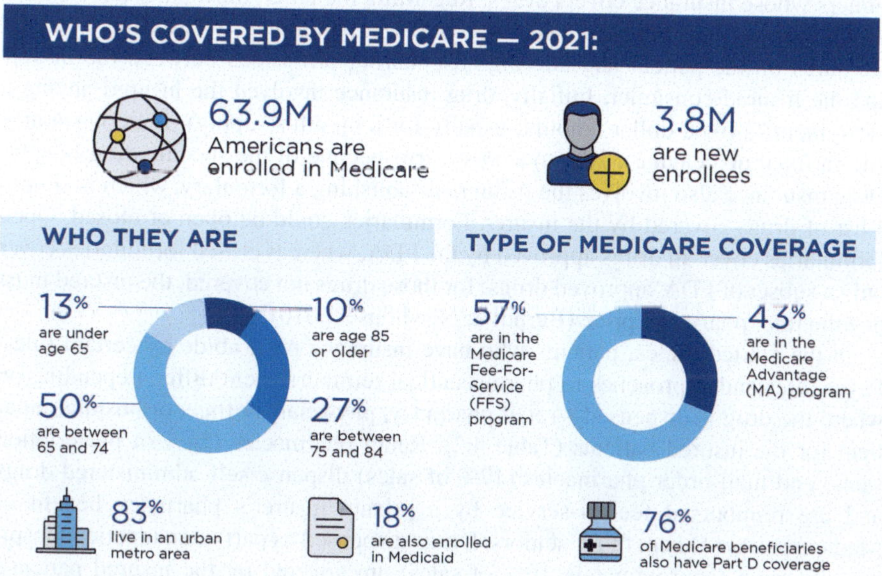

Fig. 4.4 Medicare beneficiaries in 2021. (CMS, 2024a)

prescription drug coverage was not added to the Medicare program earlier than 2003. First, prescription drug coverage has usually been tied to the fate of broader proposals for Medicare reform. Second, action has been hampered by divided government, federal budget deficits, and ideological conflict between those seeking to expand the traditional Medicare program and those preferring a more significant role for private healthcare companies. Third, the provisions for the Medicare Prescription, Drug, Improvement, and Modernization Act of 2003 reflect earlier missed opportunities. Policymakers concluded from past episodes that participation in the new program should be voluntary, with Medicare beneficiaries and taxpayers sharing the costs (Oliver et al., 2004). The result is that historically, Medicare was forbidden from negotiating drug prices. This changed with the passage of the Inflation Reduction Act of 2022, a topic discussed further in Sect. 8.2.1.

4.2.2 *Pricing and Reimbursement of Pharmaceuticals*

Pharmaceutical pricing is a complex topic but an important one. As discussed in Sect. 3.2.4, in the United States, there is no single price for a drug that all participants know. Instead, drug pricing has many different prices depending on who buys, sells, and when and where the transaction occurs. There are two options for prices paid by consumers: (1) completely cash-paying uninsured consumers and (2) consumers whose insurance covers drugs. Regarding the latter, third-party payers have attempted to constrain rising pharmaceutical expenditures by exerting financial pressures on the patient and altering cost-sharing provisions between the insurer and the insured consumer. Initially, drug insurance involved the insured paying a copayment (a fixed dollar amount, usually for a month's supply) or a coinsurance rate (a fixed percentage of the total cost), irrespective of the specific drug bought. Drug insurance also involves the insurers establishing a formulary, which is simply a list of drugs covered by the insurer. Formularies could be open or closed. Open formularies cover all drugs approved by the FDA, whereas closed formularies cover only a subset of FDA-approved drugs; for those drugs not covered, the insured must pay the total pharmacy price (Berndt & Newhouse, 2010).

In the United States, patients who have insurance must abide by certain rules. Payer rules and approaches to pharmaceutical reimbursement differ, depending on where the drug is dispensed—retail pharmacy, physician's office, or hospital inpatient for the insured patients (Table 4.2). Retail pharmacies (54% of prescription sales) and mail-order pharmacies (17% of sales) dispense self-administered drugs and are reimbursed fee-for-service by a private insurer's pharmacy benefit or Medicare's Part D benefit for seniors. Drugs dispensed as part of an inpatient hospital admission (approximately 10% of sales) are covered by the insured patient's inpatient benefit or Medicare Part A for seniors. Inpatient drugs and all other inpatient costs are reimbursed in a single bundled payment for the hospital admission. Drugs dispensed in physicians' clinics (approximately 12% of sales), including

Table 4.2 US reimbursement rules for insured patients

		Channel		
		Retail pharmacy	Physician office	Hospital inpatient
Features	Medicine type	Orals, creams, and self-injectables	Biologics, infusions, and vaccines	All types
	Benefit	Pharmacy/Medicare D	Medical/Medicare B	Hospital/Medicare A
	Reimbursement	Tiered formularies with access controls	Buy-and-bill with average selling Price (ASP) + 6%	Hospital is paid according to the diagnosis-related group (DRG) per admission

Adapted from Danzon (2014) with permission from Elsevier

infusions and vaccines, are covered by the insured patient's medical benefit or by Medicare Part B[1] for seniors (Danzon, 2014).

In the case of Medicaid, the Average Manufacturer Price (AMP) is critical. AMP is the price wholesalers pay to purchase drugs from manufacturers. Another important metric is known as "Medicaid Best Price". Medicaid Best Price was a legislated policy solution enacted over 30 years ago to address high drug costs and make Medicaid drug spending more manageable for states. Under this policy, a drug manufacturer must offer state Medicaid programs the best price given to any other purchaser (with a few exceptions) with a mandatory rebate of 23.1% off the list price. Medicaid programs must, in turn, cover all of the manufacturer's prescription drugs, with few exceptions (Baghdadi, 2017).

4.2.3 *Statistics of Medicare Reimbursements*

CMS maintains a detailed database of the spending for all programs it administers. The Medicare Part B Spending by Drug dataset presents spending information for Medicare Part B drugs—drugs administered in doctors' offices and other outpatient settings and paid through the Medicare Part B program. Medicare Part B claims (e.g., claims from physicians, durable medical equipment and other suppliers, outpatient hospitals, etc.) contain information on drugs administered and billed directly to Medicare by providers. Drug spending metrics for Medicare Part B drugs are based on total spending, derived from summing the three revenue center payment fields on the claim referring to Medicare payment, deductible, and coinsurance. This represents the product's total value, including the Medicare payment and

[1] Many new and expensive biologics are physician-dispensed, including many oncologic drugs. Therefore, the reimbursement rules for insured patient's medical benefit (or Medicare Part B for seniors) are critical to the pricing of biologics in the US (Danzon, 2014).

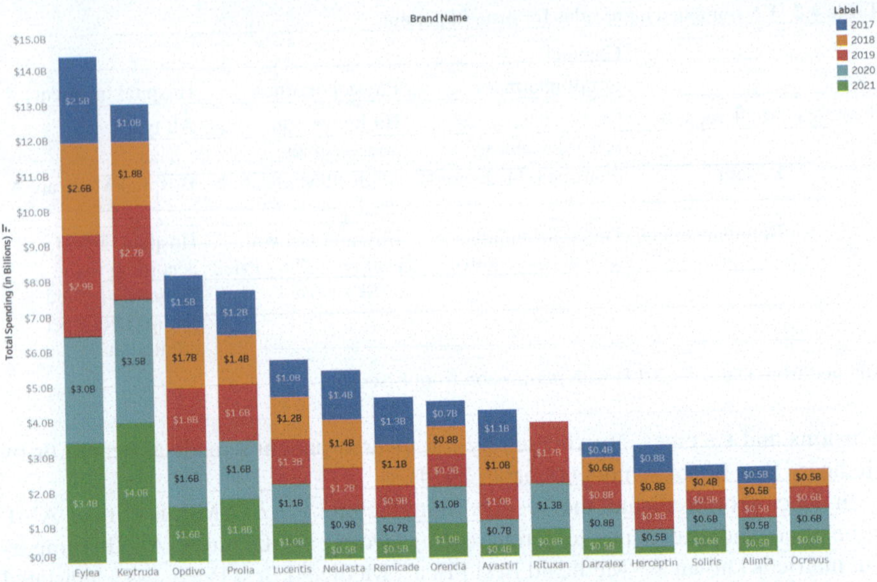

Fig. 4.5 Total Medicare Part B spending on top 15 prescription drugs (2017–2021). (Data Source: CMS, 2024b)

beneficiary liability. CMS' Part B drug spending metrics are calculated at the Healthcare Common Procedure Coding System (HCPCS) level (CMS, 2024b). The total spending on the top 15 Medicare Part B prescription drugs incurred from 2017–2021 is shown in Fig. 4.5.

The Medicare Part D Spending by Drug dataset presents spending information for Medicare Part D drugs - drugs patients generally administer themselves and that are paid through the Medicare Part D Prescription Drug program. Part D drug information is available from the Part D Prescription Drug Event (PDE) data, available for a subset of Medicare beneficiaries who enroll in Part D (approximately 77% of Medicare beneficiaries). The Medicare Part D Spending by Drug data focuses on average spending per dosage unit and changes in average spending per dosage unit over time. The data also displays spending information for the manufacturer(s) of the drugs, as well as consumer-friendly information on drug uses and clinical indications. Medicare Part D PDE records are summarized by drug by linking National Drug Codes (NDCs) available in the PDE data to a commercially available database and aggregated across all strengths, dosage forms, and routes of administration to the drug brand name and generic name (CMS, 2024b). The total spending on the top 15 Medicare Part D prescription drugs incurred from 2017–2021 is shown in Fig. 4.6.

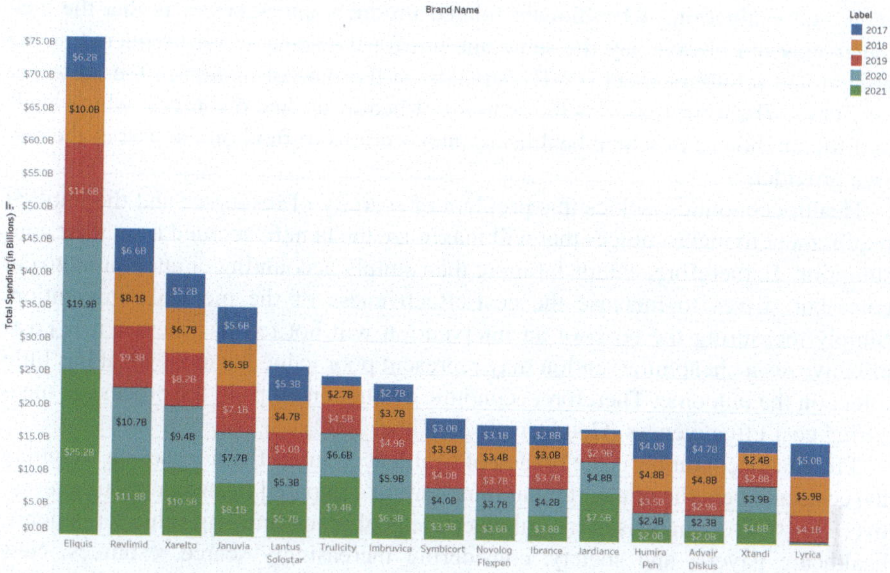

Fig. 4.6 Total Medicare Part D spending on top 15 prescription drugs (2017–2021). (Data Source: CMS, 2024b)

4.3 New Economic Evaluation Methodologies in Healthcare

Economic value is defined as how well the service or product serves its intended purpose at a price the customers are willing to pay. Therefore, in the context of economics, a methodology has to be developed that helps evaluate the economic value of a commodity by taking into account the perceived benefit of the product or service and the costs at which this benefit is realized. This methodology, known as economic evaluation, is the process of measuring cost-effectiveness. A simple example helps illustrate this concept. A significant portion of a family's annual budget is typically consumed in purchasing and maintaining a car. Therefore, buying cars is an essential decision that consumers make. With the variety of choices (in all budget ranges) available in the marketplace, customers attempt to ensure they get a good deal. One way the customers can evaluate their various options is by comparing the differences in specifications among the cars that fit their budget. Based on the freely available information to the customers, they can judge a car's performance by its engine output, fuel efficiency, accessories, and safety ratings.

Such an evaluation is difficult to do in the case of healthcare-related products and services. The first reason is that due to the specialized nature of medical diagnosis, customers do not always have control over what they can purchase. Customers must rely on the professional advice of the physician. The customer cannot test the product before consuming it, and there is an element of trust in the relationship. Therefore, medical services have a specialized operating model that is significantly different from other typical commodities. The second reason for the difficulty in economic evaluation of health-care-related products and services is that the customer may not always pay the same amount for the same service rendered as the pricing and reimbursement (P&R) depends on the type of insurance the customer may have. The third reason is the scarcity of health service resources, which leads to difficult choices of which healthcare interventions to fund on the part of the service provider.

Health economics tackles this problem of scarcity of resources and the implicit requirement to make choices that will maximize the benefit accrued from their consumption. It, therefore, entails far more than simply accounting or attempting to cut costs but strives to increase the cost-effectiveness of the medical intervention. Simply measuring the costs of an intervention will not tell us whether it is cost-effective, as a cheap intervention may represent poor value for money if it has little effect on the outcome. Therefore, economic evaluation helps in the process of measuring cost-effectiveness (Goodacre & McCabe, 2002).

Pharmacoeconomics is the branch of health economics that focuses on weighing the costs and benefits of a particular intervention compared to an analogous alternative. This type of analysis is critical given the goal of maximizing value for patients, healthcare payers, and society, considering increasingly scarce resources. New healthcare interventions (drugs, medical devices, or services) are generally costlier than the existing ones. Nevertheless, these usually provide added benefits (or 'increased value') over the standard of care. Thus, decision-makers (e.g., healthcare professionals, politicians, and other stakeholders) have to consider whether or not the new intervention is affordable and efficiently uses limited resources. Complete pharmacoeconomic evaluations also inform reimbursement or coverage decisions, which can ultimately result in price negotiations (Tonin et al., 2021). An economic evaluation in healthcare will measure two parameters—cost and outcome (effect). Several types of economic evaluation are recognized, as shown in Fig. 4.7. The measurement of outcome determines what type of economic evaluation has been performed.

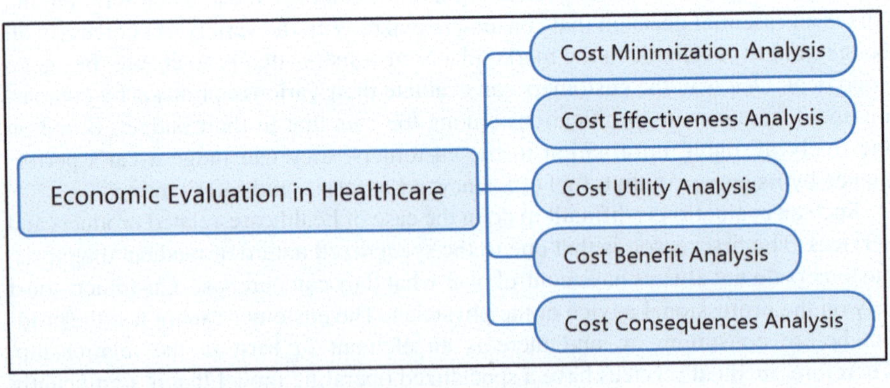

Fig. 4.7 Various types of economic evaluations in healthcare

- ***Cost Minimization Analysis (CMA):*** CMA comprises the least costly alternatives when the outcomes of two or more therapies are virtually identical. CMA involves calculating drug costs to analyze the least costly drug or therapeutic modality. It also reflects the cost of preparing and administering a dose. This cost evaluation method is used most often in evaluating the cost of a specific drug. This method can only be used to compare two products that are equivalent in dose and therapeutic effect. Therefore, this method is most helpful in comparing generic and therapeutic equivalent drugs. In many cases, there is no reliable equivalence between two products, and if therapeutic equivalence cannot be demonstrated, then the cost-minimization analysis is inappropriate (Rai & Goyal, 2018).
- ***Cost Effectiveness Analysis (CEA):*** For a cost-effectiveness analysis, the outcomes of the alternative strategies are not equivalent and are measured in unidirectional natural units, such as lives saved or change in pain score. The results are, therefore, helpful in determining technical efficiency (Goodacre & McCabe, 2002).
- ***Cost-Utility Analysis (CUA):*** For a cost-utility analysis, the outcomes of healthcare interventions are measured in units of health outcome that combine quality and quantity of life and can thus be compared between different interventions and health problems. The most well-known example of a measure of health utility is the quality-adjusted life year or QALY. Another measure of health utility is disability-adjusted life year (DALY). Calculating QALYs entails measuring the quality of life on a scale from zero to one, where zero equates to death, and one equates to perfect health. The period (in years) this quality assessment applies is multiplied by its quality weighting to give the number of QALYs (Goodacre & McCabe, 2002). An illustration of QALYs gained from treatment is shown in Fig. 4.8. This demonstrates the QALYs that can be gained by an individual from receiving treatment as opposed to no treatment. The area under the curve equates to the total QALY value (Whitehead & Ali, 2010).
- ***Cost Benefit Analysis (CBA):*** CBA is an economic analysis in which both the costs and consequences are expressed in monetary terms. The numerator states the monetary benefit gained from the treatment, and the denominator defines the monetary investment for the treatment. CBA comes into play to value both incremental costs and outcomes in monetary terms, allowing a direct calculation of the net monetary cost of achieving a health outcome. A gain in life-years may be regarded as the cost of that life-year's productive value to society. The methods for analyzing gains in quality of life include techniques such as willingness-to-pay, where the amount individuals would be willing to pay for a quality-of-life benefit is assessed (Rai & Goyal, 2018).
- ***Cost Consequences Analysis (CCA):*** Cost-consequence analysis is a form of comparative economic analysis that evaluates two or more policy alternatives in terms of their relative costs and outcomes, where the outcomes are not summarized in a single measure, and multiple outcomes of interest are reported (Turner et al., 2021). The limitation of this type of analysis is that it does not allow a transparent assessment of whether the health gained from the expenditure of

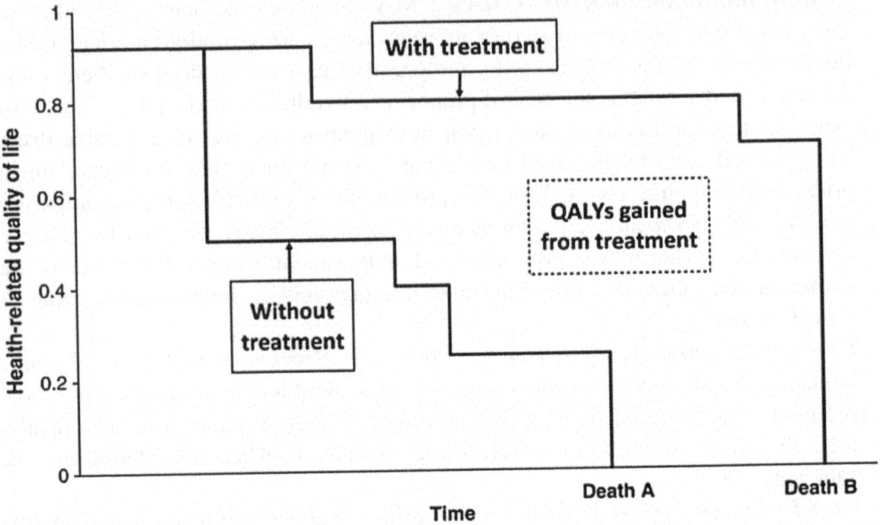

Fig. 4.8 QALYs gained from treatment. (Whitehead & Ali, 2010) (Reprinted with permission from Oxford University Press)

limited resources is being maximized. A decision maker can only be confident of this when one intervention dominates the others on all outcomes and costs (Goodacre & McCabe, 2002).

4.3.1 Application of Pharmacoeconomics in Decision-Making

The first and most crucial step in the assessment of a health intervention is to measure the costs and benefits of that intervention. The economic evaluation methodologies discussed in the previous section are designed to help with the decision-making process when evaluating the value of a new health intervention when compared to the existing standard of care (SOC). The goal of any cost-effectiveness study is to compare multiple therapies or strategies that can be used to treat the same disease or condition. Sometimes, the multiple treatment options are alike and have similar costs and benefits, and sometimes, the treatment options are entirely dissimilar. Therefore, the costs and benefits of each choice may vary tremendously, and it may not be obvious how to compare them (Bhattacharya et al., 2014).

Figure 4.9 visually represents what can happen when comparing a new health intervention to a comparator within a cost-effectiveness/cost-utility analysis. As can be expected, if the new intervention falls in Category B (less costly and more effective than SOC), then its cost-utility analysis is favorable. Conversely, if the new intervention falls into Category D (more costly and less effective than SOC), its

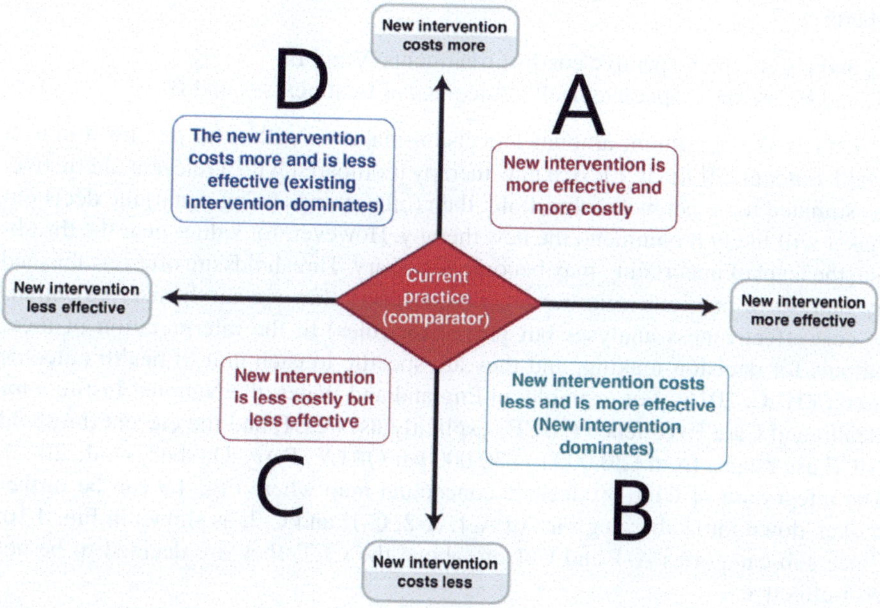

Fig. 4.9 Possible outcomes when comparing new intervention to standard-of-care. (Adapted from Turner et al. 2021)

cost-utility analysis is unfavorable, and the existing SOC will continue to dominate. If a new intervention falls in either Category A (more costly and more effective than SOC) or Category C (less costly but less effective than SOC), then a further nuanced evaluation is required as neither the new intervention nor the SOC may be the dominant treatment option. This evaluation is conducted by calculating the Incremental Cost-Effectiveness Ratio (ICER) that helps in establishing the Cost-Effectiveness Threshold (CET).

4.3.2 *ICER and CET*

ICER is a calculation that is at the heart of the cost-effectiveness analysis. ICER is the ratio of the incremental costs of pursuing one treatment over another to the incremental benefits of that treatment (Eq. 4.1). The formula for ICER is simply the ratio of the difference in costs to the difference in health outcomes. As long as neither treatment dominates the other, the ICER between the two will always be positive (Bhattacharya et al., 2014).

Equation 4.1: ICER Calculation

$$ICER_{A,B} = \frac{C_A - C_B}{E_A - E_B} > 0 \tag{4.1}$$

where:

C_A and C_B are the respective costs of treatments A and B
E_A and E_B are the respective health outcomes of treatments A and B

CET is the maximum amount a decision-maker is willing to pay for a unit of health outcome. If the ICER of a new therapy (compared with a relevant alternative) is estimated to be below the threshold, then (other things being equal) the decision-maker will likely recommend the new therapy. However, for values near the threshold, the level of uncertainty may become necessary. Thresholds are often established by analysis of previous (reimbursement) decisions: they are not themselves outputs of cost-effectiveness analyses but guides (or rules) to the interpretation of these outputs for decision-making, and they are specific to each unit of health outcome used (YHEC, 2016). For example, in England and Wales, the National Institute for Health and Care Excellence (NICE) explicitly uses CET, and the current threshold NICE use ranges from £20,000 to £30,000 per QALY (Perry-Duxbury et al., 2022). The integration of CET produces a conceptual map where Fig. 4.9 can be further broken down into sub-categories of A-1, A-2, C-1, and C-2, as shown in Fig. 4.10. Since sub-categories A-1 and C-1 are above the CET, they are deemed to be not cost-effective.

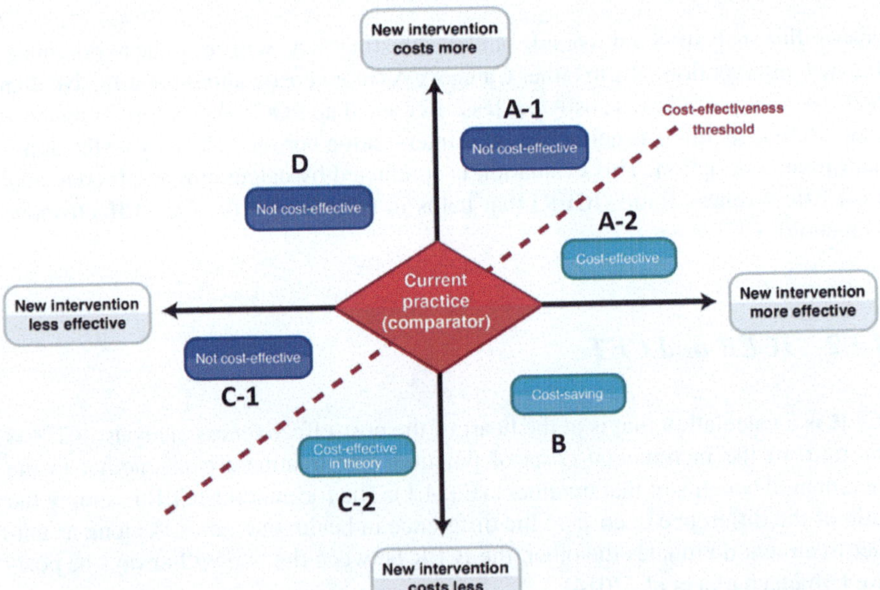

Fig. 4.10 Integration of CET with CEA outcomes map. (Adapted from Turner et al. 2021)

4.3.3 Value-Based Pricing Systems

One of the biggest criticisms of the pharmaceutical industry is that in the United States, pharmaceutical companies can set the price of a prescription drug at whatever they believe the market will bear. As a result, the prices of prescription drugs are not tied to the value those drugs provide to patients, as they are in many of our peer nations. Countries such as Germany and Australia use value-based pricing, ensuring that the price of a drug is based on the benefit provided to patients in terms of quality of life or efficacy; this pricing approach makes their drugs cheaper and more valuable to the patients taking them. Because of the pharmaceutical industry's ability to set largely unchecked prices, drug prices in the United States are much higher than in peer nations. In addition to starting at a higher price, prescription prices have risen for several decades at an unsustainable rate, higher than inflation. These high prices and price increases have a tangible impact on patients: Nearly 3 in 10 American adults reported not taking medicines as prescribed in 2019 due to cost. American taxpayers also bear the burden of high drug prices, with Medicare and Medicaid spending nearly $290 billion on prescription drugs in 2019. As a result, there has been considerable discussion at the state and federal levels on how to lower overall prescription spending and high underlying prices (Waldrop, 2021). One proposed approach to address the high prices for prescription drugs is to switch to a value-based pricing (VBP) system that would better serve patients and improve the efficiency of healthcare spending in the United States.

Recently, there has been increased interest in manufacturer and payer agreements that tie reimbursement to product performance. These are generally referred to as performance-based risk-sharing arrangements or outcomes-based agreementss (OBAs). In an OBA, performance in a defined patient population is tracked over a specified period in a defined population or at the individual patient level, and the amount or level of reimbursement is determined based on the outcomes achieved (Brown et al., 2018). These outcomes could be future measures of clinical or intermediate endpoints ultimately related to patient quality or quantity of life. The hope is that with these kinds of performance- or outcomes-based agreements, the patients will have access to novel and potentially beneficial healthcare technologies under significant uncertainty and cost pressures. Many types of health outcomes-based schemes have been proposed in the literature, and researchers have created a detailed taxonomy that characterizes vital features of the schemes (Carlson et al., 2010). This taxonomy is presented in Fig. 4.11 and further explained in Table 4.3.

4.3.4 Impact of VBP Systems on Pharmaceuticals

The developers of innovative pharmaceuticals, devices, and diagnostic tests face a complex global marketplace. First, since their products have global public goods attributes, many countries can come into play, and each has its health delivery

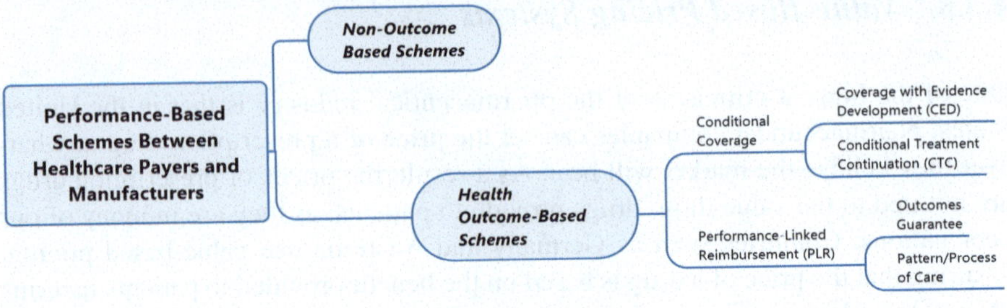

Fig. 4.11 Taxonomy of performance-based health outcomes reimbursement schemes. (Adapted from Carlson et al. 2010 with permission from Elsevier)

Table 4.3 Types of reimbursement schemes

Reimbursement Scheme	Features
Non-outcome based population level	In this scheme, the government or health insurer sets the level of public payment for a given group or cluster of drugs. Manufacturers are free to set the price of their produce above the reference prices, but the patient has to pay the difference. Drug pricing is not linked to a health outcome but is somewhat dependent on the market share and volume and could include utilization caps at the patient level.
Health outcomes-based conditional coverage	These are schemes where coverage is conditionally granted upon initiating a data collection program. Conditional coverage-based schemes can be further classified into coverage with evidence development (CED) or conditional treatment continuation (CTC). In CED, a binary coverage decision is conditioned upon collecting additional population-level evidence from a pre-specified scientific study to support continued, expanded, or withdrawn coverage. CTC implies coverage for individual patients is conditioned upon meeting short-term treatment goals (e.g., tumor response or lower cholesterol).
Health outcomes based performance-linked	These are schemes where the reimbursement level for covered products is tied to the measure of clinical outcomes in the "real world.". Performance-linked reimbursement (PLR) can be further classified into outcomes guarantee or pattern/process of care. In outcomes guarantee schemes, the manufacturer provides rebates, refunds, or price adjustments if their product fails to meet the agreed-upon outcome targets, which could be clinical or intermediate endpoints. In pattern/process of care schemes, the reimbursement level is tied to the impact on clinical decision-making or practice patterns (e.g., whether or not patients adhere to the treatment course suggested by a risk-predicting genomic test).

Adapted from Carlson et al. (2010) with permission from Elsevier

system, priorities, and approach to pricing and reimbursement. For the innovative pharmaceutical industry to be sustainable and robust, the revenues generated through the pricing and reimbursement system must at least be on par with the growth of R&D costs (Garrison & Towse, 2017). This is extremely difficult to achieve in an operating environment where the overall industry productivity in the number of new compounds has been flat over the past 11 years—averaging 30–35 new compounds per year. However, R&D costs have been rising at 8.5% per annum, and the latest estimate of the average cost per new molecule entity is $2.6 billion. If R&D costs continue to rise, medicines targeted to smaller indications will require higher prices, creating a vicious cycle where effective interventions may be consistently above the CET in the A-1 subcategory of Fig. 4.10. In value-based pricing systems, such drugs may not be favorably reviewed by decision-makers, potentially depriving the patients of their benefits.

Such a scenario creates a conundrum in the case of the pricing of treatments for ultra-rare diseases. Here, treatments, such as enzyme replacement, can cost as much as $500,000 per year. So even if they extended life at full health, it is unlikely they could ever be cost-effective when judged against conventional standards. In addition, these therapies often fulfill a previously unmet need; therefore, there is no existing therapy to be replaced, which might generate cost offsets. Therefore, if, for whatever reason, it is decided to reimburse these therapies, VBP is unlikely to help set an appropriate price since the price will be too low, either to encourage manufacturers to launch the products or, in the long run, to stimulate research into rare conditions. A second case where the VBP may be deemed inappropriate is in the pricing of 'cures', such as gene therapy, for conditions currently being treated at great expense. For example, suppose gene therapy became available for those patients on long-term hemodialysis. If the need for life-long dialysis could be obviated, such a cure could be worth millions of dollars based on cost savings from current care alone. If the new therapy offered a higher quality of care, the price implied by an ICER up to the jurisdiction's current decision-making threshold would likely be even higher. Therefore, such a treatment could be expensive if VBP were applied (Drummond & Towse, 2019).

4.4 Summary—Evolving Marketplace for Pharmaceuticals

The practical reality is that the pharmaceutical industry may need to account for a future where VBP is the norm and not the exception. In such a scenario, designers of the pharmaceutical products will routinely be asked what can be (further) done to reduce the costs so that the product does not cross the CET. While the debate around the merits of VBP continues in the literature, changes in the law are taking place. With the introduction of the Inflation Reduction Act 2022, significant savings are expected for Medicare, which serves 20% of patients. Although, the IRA does not address cost savings for the other 80% of patients in the private market. However, it can be anticipated that future changes in the law could address further cost savings.

Therefore, evaluating how the pharmaceutical industry could adapt to this potential new paradigm shift is necessary. The changes in legislation often precede changes in the marketplace. In that regard, there is no wonder that the pharmaceutical marketplace will also evolve to react to the realities of the law. ***Could an evaluation of the pharmaceutical drug development lifecycle help identify opportunities for enhancing the value proposition of our products?***

References

Baghdadi, R. (2017). Health policy brief: Medicaid best price. *Health Affairs*. https://doi.org/10.1377/hpb2017.8

Berndt, E. R., & Newhouse, J. P. (2010). *Pricing and reimbursement in US pharmaceutical markets*. Retrieved from https://www.nber.org/system/files/working_papers/w16297/w16297.pdf

Bhattacharya, J., Hyde, T., & Tu, P. (2014). *Health economics*. Palgrave Macmillan.

Britannica. (2024a). Tulip. In *Encyclopedia Britannica*.

Britannica. (2024b). Tulip Mania. In *Encyclopedia Britannica*.

Brown, J. D., Sheer, R., Pasquale, M., Sudharshan, L., Axelsen, K., Subedi, P., Wiederkehr, D., Brownfield, F., & Kamal-Bahl, S. (2018). Payer and pharmaceutical manufacturer considerations for outcomes-based agreements in the United States. *Value in Health, 21*(1), 33–40. https://doi.org/10.1016/j.jval.2017.07.009

Carlson, J. J., Sullivan, S. D., Garrison, L. P., Neumann, P. J., & Veenstra, D. L. (2010). Linking payment to health outcomes: A taxonomy and examination of performance-based reimbursement schemes between healthcare payers and manufacturers. *Health Policy, 96*(3), 179–190. https://doi.org/10.1016/j.healthpol.2010.02.005

CMS. (2024a). *Medicare beneficiaries at a glance (2023 Edition)*. Retrieved from https://data.cms.gov/sites/default/files/2023-09/Medicare%20Beneficiaries%20at%20a%20Glance%20DY2021.pdf

CMS. (2024b). *Medicare part B and D spending by drug*. https://data.cms.gov/

Danzon, P. M. (2014). Pricing and reimbursement of biopharmaceuticals and medical devices in the USA. In A. J. Culyer (Ed.), *Encyclopedia of health economics* (pp. 127–135). Elsevier. https://doi.org/10.1016/B978-0-12-375678-7.01209-8

Drummond, M., & Towse, A. (2019). Is rate of return pricing a useful approach when value-based pricing is not appropriate? *The European Journal of Health Economics, 20*(7), 945–948. https://doi.org/10.1007/s10198-019-01032-7

Garrison, L. P., & Towse, A. (2017). Value-based pricing and reimbursement in personalised healthcare: Introduction to the basic health economics. *Journal of Personalized Medicine, 7*(3), 10. https://www.mdpi.com/2075-4426/7/3/10

Girdzijauskas, S., Štreimikiene, D., Čepinskis, J., Moskaliova, V., Jurkonyte, E., & Mackevičius, R. (2009). Formation of economic bubbles: Causes and possible preventions. *Ukio Technologinis ir Ekonominis Vystymas, 15*(2), 267–280. https://doi.org/10.3846/1392-8619.2009.15.267-280

Goodacre, S., & McCabe, C. (2002). An introduction to economic evaluation. *Emergency Medicine Journal, 19*(3), 198–201. https://doi.org/10.1136/emj.19.3.198

Krajewski, L. J., Ritzman, L. P., & Malhotra, M. K. (2010). *Operations management: Processes and supply chains* (9th ed.). Pearson Education, Inc.

McClure, J. E., & Thomas, D. C. (2017). Explaining the timing of tulipmania's boom and bust: Historical context, sequestered capital and market signals. *Financial History Review, 24*(2), 121–141. https://doi.org/10.1017/S0968565017000154

Oliver, T. R., Lee, P. R., & Lipton, H. L. (2004). A political history of medicare and prescription drug coverage. *The Milbank Quarterly, 82*(2), 283–354. https://doi.org/10.1111/j.0887-378X.2004.00311.x

Perry-Duxbury, M., Lomas, J., Asaria, M., & van Baal, P. (2022). The relevance of including future healthcare costs in cost-effectiveness threshold calculations for the UK NHS. *PharmacoEconomics, 40*(2), 233–239. https://doi.org/10.1007/s40273-021-01090-x

Popa, C., Holvoet, K., Van Montfort, T., Groeneveld, F., & Simoens, S. (2018). Risk-return analysis of the biopharmaceutical industry as compared to other industries. *Frontiers in Pharmacology, 9*, 1108. https://doi.org/10.3389/fphar.2018.01108

Rai, M., & Goyal, R. (2018). Chapter 33—Pharmacoeconomics in healthcare. In D. Vohora & G. Singh (Eds.), *Pharmaceutical medicine and translational clinical research* (pp. 465–472). Academic. https://doi.org/10.1016/B978-0-12-802103-3.00034-1

Ruffin, R. J., & Gregory, P. R. (2001). *Principles of economics* (7th ed.). Addison Wesley Longman, Inc.

Tonin, F. S., Aznar-Lou, I., Pontinha, V. M., Pontarolo, R., & Fernandez-Llimos, F. (2021). Principles of pharmacoeconomic analysis: The case of pharmacist-led interventions. *Pharm Pract (Granada), 19*(1), 2302. https://doi.org/10.18549/PharmPract.2021.1.2302

Turner, H. C., Archer, R. A., Downey, L. E., Isaranuwatchai, W., Chalkidou, K., Jit, M., & Teerawattananon, Y. (2021). An introduction to the Main types of economic evaluations used for informing priority setting and resource allocation in healthcare: Key features, uses, and limitations [review]. *Frontiers in Public Health, 9*. https://doi.org/10.3389/fpubh.2021.722927

Waldrop, T. (2021). *Value-based pricing of prescription drugs benefits patients and promotes innovation.* https://www.americanprogress.org/article/value-based-pricing-prescription-drugs-benefits-patients-promotes-innovation/#:~:text=Value%2Dbased%20pricing%20has%20the%20potential%20to%20effect%20lower%20health,and%20create%20better%20patient%20outcomes

Whitehead, S. J., & Ali, S. (2010). Health outcomes in economic evaluation: The QALY and utilities. *British Medical Bulletin, 96*(1), 5–21. https://doi.org/10.1093/bmb/ldq033

Wolff, M. (2022). *The production possibilities curve in economics.* https://articles.outlier.org/ppc-curve

YHEC. (2016). Cost-effectiveness threshold. *York Health Economics Consortium.* Retrieved October 14 from https://yhec.co.uk/glossary/cost-effectiveness-threshold/

Chapter 5
Evaluation of Pharmaceutical Drug Development Lifecycle

Success is not final; failure is not fatal. It is the courage to continue that counts.

–Winston Churchill

Contents

Abstract Value in healthcare is the measured improvement in a patient's health outcomes for the cost of achieving that improvement. The goal of value-based care transformation is to enable the healthcare system to create more value for patients. The pharmaceutical industry is very conscious of any development efforts that can help reduce costs and minimize the time it takes for the products to reach the commercial market. Similarly, numerous opportunities exist to help enhance the value

B. Mittal, *Sustainable Global Health Systems and Pharmaceutical Development*, AAPS Advances in the Pharmaceutical Sciences Series 60,
https://doi.org/10.1007/978-3-031-50415-0_5

proposition of pharmaceuticals. Among all the opportunities, product design, development, and manufacturing play a significant role. A shining example of the impact of pharmaceuticals on the society was witnessed recently. The year 2020 has been challenging in all socioeconomic aspects. Not only had the COVID-19 pandemic devastated the global economy, but it also bared the fault lines that continue to exist in society. Our combined fragility has never been as evident as these current times. On a positive note, though, this pandemic has also allowed us to reset our thinking about global health systems and sincerely evaluate their sustainability. The speed by which the mRNA vaccines were developed and introduced globally is something we in the pharmaceutical industry should be very proud of. The mRNA vaccines represent a promising alternative to conventional vaccine approaches because of their high potency, capacity for rapid development, and potential for low-cost manufacture and safe administration. Recent technological advances have largely overcome these issues, and multiple mRNA vaccine platforms against infectious diseases and several types of cancer have demonstrated encouraging results in both animal models and humans. However, other aspects of drug development, such as discovery, clinical trials, manufacturing, and regulation, have also seen significant positive changes in the last few years. For example, the philosophy of patient-focused drug development (PFDD) and modernization of pharmaceutical manufacturing provides a new template for product development that hopes to reduce drug development time and make it more cost-effective. Therefore, a review of the current state of the pharmaceutical industry is important as the battle against disease is far from being won.

Keywords Continuous innovation · Growth Mindset · Product lifecycle · Rational drug design · Clinical trial design · Quality by design · Patient-centric dosage design · Medication adherence · New product launch performance · Manufacturing issues · Marketing exclusivity · Product line extensions

5.1 Introduction

Pharmaceuticals is a harsh industry. Not only is its mandate to develop and commercialize life-saving medicines a lofty one but so is its unbridled diligence to continue to innovate and deliver day after day despite high rates of failures. Irrespective of all its contributions, the pharmaceutical industry is sometimes subjected to disproportionate and extreme criticism. Many times, it has had to justify why it takes so long to discover new drugs, why they cost so much, and why their productivity is so low. Critics also assert that poor R&D strategies have led to severe pipeline issues and that R&D has been based on market considerations as opposed to 'real health needs,' resulting in a plethora of ambivalent 'me-too' medicines, and that many essential disease areas and patient groups are being neglected. Unfortunately, such accusations unfairly belittle the work of numerous scientists, engineers, and researchers passionate about bringing new therapies to market. As discussed in

Chap. 3, the pharmaceutical industry has a business model based on a high level of financial investment without much guarantee of success. There have been other industries that have had similar criticisms in the past, and not only were they able to win over critics, but they also set up new benchmarks that are examples for all of us to follow today. One such industry is that of aeronautics and aerospace.

5.1.1 Philosophical Digression: Continuous Innovation with a Progressive Growth Mindset

National Aeronautics and Space Administration (NASA) is the foremost leader in space exploration. With a diverse workforce of 18,000 professionals, NASA works with many more partners to explore, discover, and expand knowledge for humanity's benefit. With an annual budget of $23.2 billion in Fiscal Year 2021, less than 0.5% of the overall US federal budget, NASA supports more than 312,000 US-based jobs, generating more than $64.3 billion in total economic output (Fiscal Year 2019). At its 20 centers and facilities across the country—and the only national laboratory in space—NASA studies Earth, including its climate, sun, solar system, and beyond. In addition to significant missions, the agency shares what it learns so that its information can make life better for people worldwide (NASA, 2022). As everyone knows, NASA is a product of the arms race that followed after World War II. However, the current lofty status of NASA was never a guarantee.

In an excellent article in Harvard Business Review on Change Management, researchers argue that the NASA of today is a very different beast from the NASA of the 1960s. Though many would call that decade NASA's golden age, these researchers argue that NASA's innovation and influence are even more significant today. During the Apollo program, NASA's funding peaked at 4.5% of the federal budget (US$5.25 billion) in 1966. However, their budget plummeted shortly after the first moon landing in 1969. Meanwhile, the organization's mission aspirations grew bolder. The Commercial Space Launch Act of 1984 (and amendments in 1988 and 2004) called for NASA to support the development and growth of commercial space. This not only necessitated a more collaborative way of operating but also called for the agency to differentiate its missions. As commercial space organizations began to drive more low-earth orbit (up to 2000 km from the Earth) activities, NASA had to shift its efforts toward exploring deep space, accomplishing further manned missions, and setting up permanent facilities on the moon as a gateway for missions to Mars and beyond. The emergence of commercial space has also gone hand in hand with an accelerating pace of technological development. This means that the technologies necessary for successful spacefaring cannot all be developed by a single organization. NASA has had to become more outward-looking and network-oriented to develop and acquire the necessary technologies. The reinvention of NASA has been an evolution spanning three phases. Each focuses on achieving different goals and is characterized by particular technology strategies, cultural values, and ways of working with external parties.

- **Traditional Model (1960s–1993)**: In its early days, NASA served as the prime contractor and the exclusive customer of space technologies. This model made sense for a few reasons. First, the frontier technologies that NASA needed were unavailable on the market; they had to be developed from scratch by specialized contractors. Second, the Cold War meant that NASA needed to have control of the resulting technology rather than the technology being made available on the market by contractors after development. Third, NASA's military heritage, and more broadly, the procurement processes of government agencies at the time, meant it operated by issuing cost-plus contracts and owning the resulting technology. In this model, the relational approach comprised positional authority and hierarchy.
- **Transitional Model (1993–2006):** In 1993, NASA was directed by the White House to collaborate with other national agencies (such as those from the European Union, Japan, Canada, and Russia) on the design and construction of the International Space Station (ISS). The ISS was a necessary step in launching human missions in deep space. NASA needed to understand what would happen to the human body during extended missions in space, and the ISS would provide a way to gather this knowledge by stationing astronauts on the space station for long periods. Nevertheless, ISS would be too expensive for any country to build alone. During this transitional phase, NASA learned how to function collaboratively within a cluster of partners. This demanded shifts in cultural values, relational approaches, and technology strategy. Further, greater cost consciousness developed as the American public and politicians began questioning the resources needed by the agency. Unlike the Apollo era, when funds were no barrier in the effort to win the space race, NASA now had to accomplish its missions as efficiently as possible and be more explicit about its added value to society. It invested more in external communication. Relationally, the hierarchical pecking order of the traditional model had to accommodate a cluster of international governmental organizations and distribute the work for efficiency. NASA's technology strategy also evolved to leverage the investments of state partners and share technical responsibility. The agency worked with its partners on developing shared technical interfaces, standards, and protocols, learning how to operate within public-public partnerships.
- **Network Model (2006–present):** This phase of NASA's evolution began with the Commercial Resupply Services program launched in 2006 to carry cargo to the ISS after the space shuttle was retired. This meant that NASA had to find ways to resupply the ISS that would be reliable and require fewer resources. This and many different factors—budget pressures, government interest in promoting commercial space, and rapidly growing commercial expertise—led NASA to seek suitable partners in the open market to deliver frontier capabilities at a lower cost than what NASA could develop them for. In 2008, NASA awarded Space X and Orbital Sciences contracts to transport cargo to the ISS.[1] This model has

[1] Space X carried out its first resupply mission in 2012 and Orbital Sciences in 2013.

changed NASA's technology strategy, which now involves fixed-price contracts within public-private partnerships. NASA does not own the resulting technology exclusively, and the commercial partners can sell their services/technology to other customers. Costs are shared, and NASA pays for milestones reached. The commercial resupply program has taught NASA how to work effectively with the commercial sector and manage ongoing public-private partnerships. Culturally, this has led to a more outward-looking agency that recognizes the innovative capacity of the market. Dealing with non-state actors has also enabled NASA to focus on accomplishing things as efficiently as possible while being cost-conscious. In this networked model, NASA has also embraced open innovation. The agency now poses innovation challenges online in open competitions, crowdsourcing solutions, and ideas as a complement to internal innovation efforts.

Over the past few decades, not only has NASA delivered crucial technologies for society, such as water filtration systems, satellite-based search-and-rescue, and UV coating on eyeglasses, but it has also evolved its dominant logic and business model. NASA has moved from a hierarchical, closed system that develops its technologies internally to an open network organization that embraces innovation, agility, and collaboration. This reinvention demonstrates that substantial organizational change is possible, even amid barriers such as regulations and politics. It offers an example of "strategic agility," or the ability to effectively (and continually) adapt how a firm operates and competes. This is not driven by a single leader but by many champions scattered around the organization who push forward initiatives that slowly create change. Given the recent directive from the US government requiring NASA to return to manned missions to the moon and beyond, the organization's new capabilities and structure are becoming even more critical. NASA can now use innovations wherever they emerge within its networks to accomplish goals such as deep space exploration, the search for extraterrestrial life, and a manned journey to Mars (Heracleous et al., 2018)

5.1.2 Opportunity for Pharmaceutical Scientists: Embrace the Advent of Networked Innovation

NASA and pharmaceuticals have a similar evolution history. Both are highly research-oriented institutions with storied histories. The products produced by these industries have benefited humanity immensely. Throughout humankind, efforts have been made to extend patients' quality of life. The United States is recognized as the global leader in biopharmaceutical innovation. That was not always the case, however. As one ponders the history of pharmaceuticals, one cannot help but wonder how the pharmaceutical industry became one of the most profitable and research-intensive industries in the United States. Like other industries, pharmaceuticals benefitted from many late nineteenth-century changes that characterized the

Industrial Revolution, such as increasing urbanization that created new markets for consumer goods, the development of some labor-saving technologies like tablet production, and especially the expansion of transportation routes with rail lines. All of this proceeded, of course, in the absence of much in the way of regulatory controls (Swann, 1995). The institutionalization of science and research in drug manufacturing, which began in the 1880s, ushered in the modern pharmaceutical industry. The two World Wars accelerated the rate of investments. Annual spending on research for the American pharmaceutical industry reached about $50 million by the early 1950s and quadrupled by the 1960s. The pharmaceutical industry fared well in comparing research expenditures among many other industries.

The 1995–2005 period was the quintessential blockbuster drugs era for big pharma, so much so that two of the largest mergers in the industry were primarily driven by single blockbusters: Lipitor® in the case of Pfizer-Warner Lambert and Celebrex® in the case of Pfizer- Pharmacia. Some of the biggest-selling drugs in the industry's history—Lipitor®, Plavix®, Nexium®, Abilify®, Seroquel®, Diovan®, and Crestor®, among others—were launched during this period. Further, most of the top-selling drugs during 1995–2005 were primary care, small-molecule therapies. During this period, primary care therapy areas accounted for nearly 80% of revenues for most big pharma portfolios (Gautam & Pan, 2016). However, after the 2008 global recession, big pharma started to embrace a 'leaner and focused' model by divesting non-core assets and focusing on their areas of strength. Some examples are Abbott split into two parts [an innovative business (AbbVie) and a diversified healthcare company (Abbott)]; GSK and Novartis swapped their oncology, consumer health, and vaccines business to create focused organizations with GSK increasing their focus on consumer health and vaccines, and Novartis on oncology. Over the past decade, big pharma has progressively tailored their pipelines to specialty medicines and biologics targeting high unmet medical needs. This trend is driven by several factors, such as better insight into the underlying disease biology to develop targeted medicines; science and technology innovation for biologics; personalized medicines and companion diagnostics; favorable regulatory framework and R&D timelines for such drugs; and pricing and reimbursement (Gautam & Pan, 2016). The result is that the industry saw a biotech boom during 2013–2016.

However, the past tells a compelling story for the future, in any case. As the United States economy continues to move forward, the pharmaceutical industry must innovate to weather the pressures that come from the maturation of markets, changes in the geopolitical landscape, regulatory hurdles, natural disasters, and any unforeseen challenges. History tells us that a financial boom cannot be sustained forever, and once it is over, it takes decades to recover. Consequently, for the healthcare sector to continue at a robust pace in the future, its pharmaceutical arm must consider the economic challenges it faces and learn to embrace them without compromising on its core mission—to ensure the well-being of humankind. This is where pharmaceuticals can evolve in their thinking, similar to how NASA evolved in 2006. As one can imagine, due to factors beyond anyone's control, all pharmaceutical companies face increasing pressures to cut costs and optimize their resources. All the macroscopic trends are changing how product design is approached. Concepts borrowed from other industries are increasingly applied to pharmaceuticals, and the field is ever-evolving. For example, product design management (PDM) is becoming an increasingly strategic concept. Researchers have highlighted four product design management activities: strengthening the relationships between firm members and the outside, fostering relationships between areas or people within the product design process, increasing information flow to a maximum, and promoting balanced heterogeneous participation in design decision-making (Chiva-Gomez, 2004). Designers of pharmaceutical products need to imbibe this new thinking into their plans.

5.2 Enhancing Value Proposition in Pharmaceutical Product Development

The pharmaceutical industry has thrived in bringing new products to the commercial market. These new products help numerous patients suffering from various ailments get a better outlook on life and, in some cases, increase their lifespan. Therapeutic drugs can be classified into two broad categories: (a) small molecules and (b) biologics. The former refers to chemically synthesized drugs, while the latter refers to products that involve components extracted from or produced by a living organism. Biologics include monoclonal antibodies (mAbs), vaccines, blood products, and advanced therapy medicinal products (ATMPs). Figure 5.1 illustrates the drug categories considered here. Each of these products is characterized by crucial specifications and/or formulations that dominate decisions related to its manufacturing and supply chain (Sarkis et al., 2021). Significant differences in the two pharmaceutical product categories are highlighted in Table 5.1.

In a report published by the FDA (FDA, 2018a), it was documented that in 2017, the Center for Drug Evaluation and Research (CDER) approved 46 novel drugs, either as new molecular entities (NMEs[2]) under Type 1 NDAs or as new therapeutic biologics under Biologics License Applications (BLAs) (Fig. 5.2). This is an encouraging trend for continued pharmaceutical innovation, but from a societal perspective, the overall number of approved novel drugs remains at an average of 32 per year. Another interesting aspect of the 46 novel drugs that were approved in 2017 is an evaluation of their route of administration, as plotted in Fig. 5.3. From an innovation perspective, these are incredible statistics as each new drug approval allows for improved quality of life and relief for patients suffering from various

[2]The terms New Molecular Entity (NME) and New Chemical Entity (NCE) are sometimes used interchangeably. However, they are distinct. An NCE is defined in 21 CFR 314.108(a) as "a drug that contains no active moiety that has been approved by the FDA in any other application submitted under 505(b) of the FD&C Act." On the other hand, NME is not defined in the statute or regulations. An NME is an active ingredient that contains no active moiety that has been previously approved by the Agency in an application submitted under section 505 of the Act or has been previously marketed as a drug in USA.

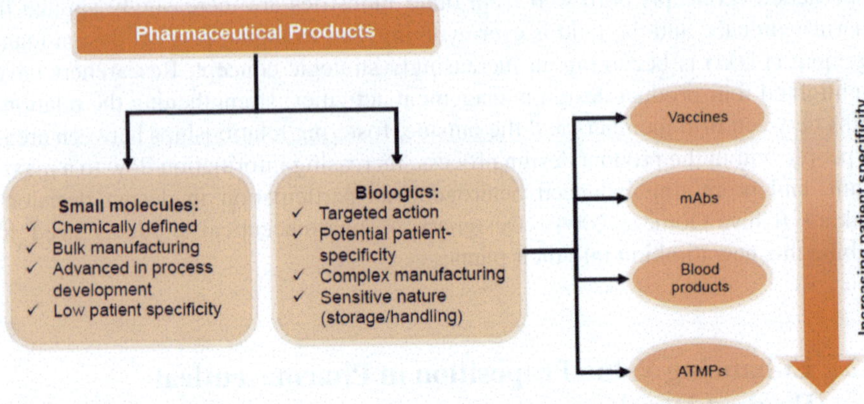

Fig. 5.1 Schematic of simplified pharmaceutical product categories. (Sarkis et al., 2021)

Table 5.1 Major differences between small molecules and biologics

Small molecules	Biologics
Produced by chemical process	Produced by living cell cultures
Low molecular weight	High molecular weight
Well-defined structure	Complex, heterogeneous structure
Mostly process-independent	Strongly process-dependent
Completely characterizable	Not entirely characterizable
Stable	Unstable
Non-immunogenic	Immunogenic

Adapted from Makurvet (2021)

ailments. However, based on the discussion in Sect. 5.1.2, one can imagine that the value proposition of pharmaceutical products will continue to be under scrutiny as the costs of medications keep going up with time. In that regard, cost optimization is an important element that needs to be designed into future pharmaceutical products.

Cost optimization is a business-focused, continuous discipline to drive spending and cost reduction while maximizing business value. As in pharmaceuticals, cost optimization is a key factor of competitiveness in numerous other manufacturing industries, such as automobiles, semiconductors, fine chemicals, etc. Manufacturers are increasingly requested to deliver faster, improve quality, and, of course, reduce costs. The need to hold these three requirements together results in the fact that cutting costs is impossible. It is necessary to optimize them (CADLOC, 2023). A strategic approach to cost optimization can bring greater and more lasting benefits supporting current and long-term business goals. Rather than short-term adaptive actions, strategic cost optimization involves continuous efforts to improve business value and build resilience.

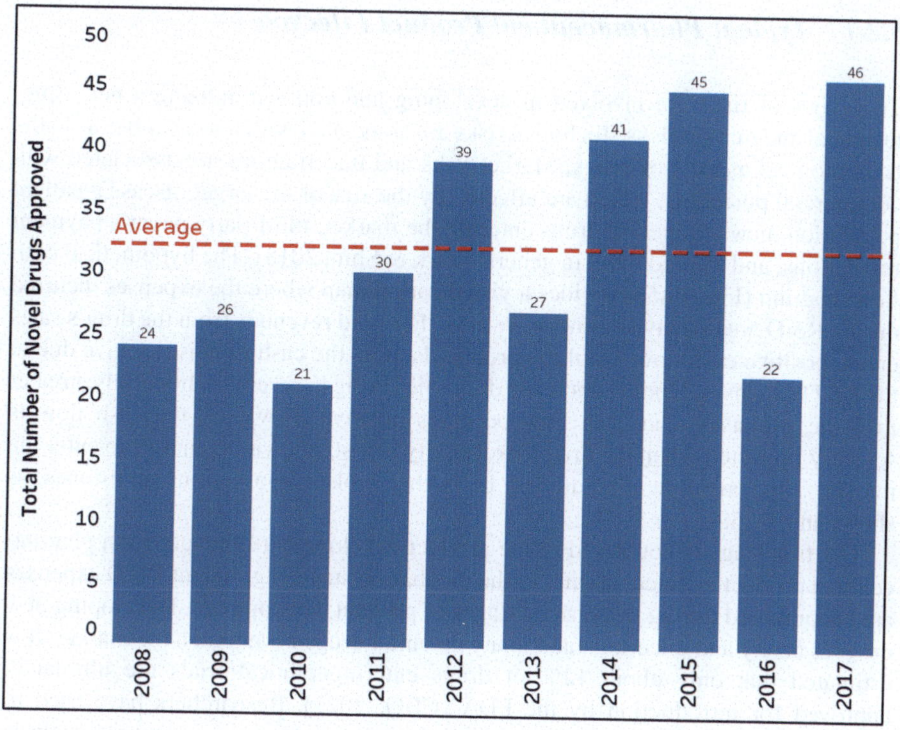

Fig. 5.2 Total number of novel drugs approved by CDER. (Mittal, 2019) (Reprinted with permission from Elsevier)

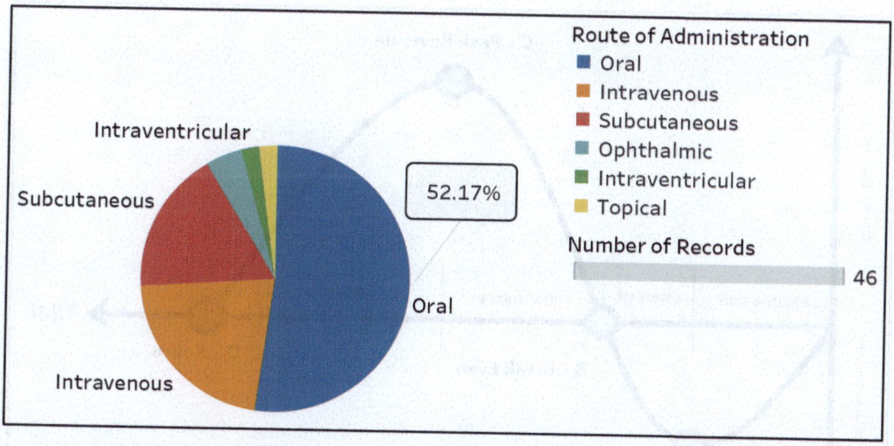

Fig. 5.3 Overview of novel drugs in 2017 by route of administration. (Mittal, 2019) (Reprinted with permission from Elsevier)

5.2.1 Typical Pharmaceutical Product Lifecycle

Two types of risks are involved in developing and commercializing a new drug: technical and market risks. Technical risks are associated with uncertainties as to the technological aspects of drugs. Market risks and uncertainties are associated with commercial potentials, which are affected by the size of the target disease's patient population, new competing drugs entering the market, third-party payer's payment regulations, and the economy in general (Park & Shin, 2018). The hypothetical cash flow diagram (Fig. 3.15) provides a visualization map where the expenses incurred during R&D are displayed as negative cash flow, and revenues from the drug's sales create positive cash flow. Until the product launch, the cash flow is negative due to the R&D expenses. Upon launch, the drug sales bring in revenues, hopefully, greater than the expenses, and cash flow becomes positive. However, the cash flow is severely impacted when the drug's exclusivity is lost, allowing generics to enter the market. This cash flow diagram can be further broken into various milestones, as shown in Fig. 5.4.

The first stage encountered in the product development lifecycle for a pharmaceutical product is "Incubation". In the incubation stage, significant R&D expenses are encountered during the various stages of product development. Developing new drugs is costly and uncertain, and many potential drugs never reach the market. It is estimated that only about 12% of drugs entering clinical trials are ultimately approved for introduction by the FDA (CBO, 2021). Researchers have tried to determine the R&D expenses during product development. In a recent analysis

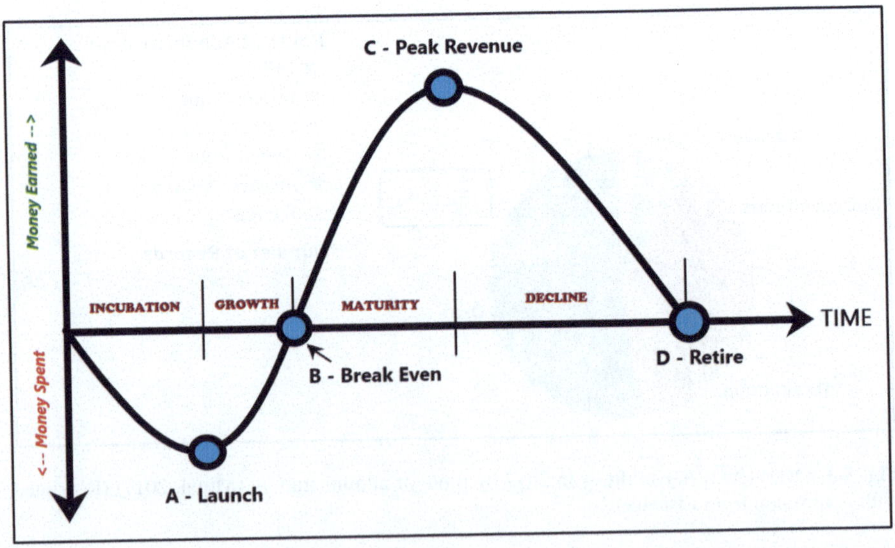

Fig. 5.4 Typical cash flow over time and milestones for pharmaceutical companies. (Adapted from Samanen (2013) and CADLOG (2023))

conducted by researchers using publicly available data, based on data for 63 therapeutic agents developed by 47 companies between 2009 and 2018, the median research and development investment required to bring a new drug to market was estimated to be $1142 million, and the mean was estimated to be $1559 million. Estimates differed across therapeutic areas, with costs of developing cancer drugs the highest. The results included the costs of failed clinical trials and varied sensitivity analyses using different estimates of trial success, preclinical expenditures, and cost of capital (Wouters et al., 2020).

The product is launched in the second stage of "Growth", and the pharmaceutical company can start generating revenues. A company's expectations about the revenues it could earn from a drug depend on the prices it anticipates the drug could command in various markets worldwide and the quantities it anticipates might be purchased at those prices. Those expectations are informed by the prices and sales volumes observed for existing drugs in various markets. The growth stage continues until a breakeven point, where revenues equal the financial investment made during product development. Beyond the breakeven point, the product enters a "Maturity" stage.

In the third stage of "Maturity", a product is assumed to be generating steadily growing revenues till it reaches peak revenue. Peak revenue is a vital metric investors frequently use to assess pharmaceutical firms' value. Peak revenues can be characterized by the height-of-peak-revenue and the time-to-peak-revenue. Both metrics are closely related to new product performance, such as cumulative sales. Intuitively, a brand with higher peak revenues will likely have higher average sales over its life cycle. As a result, cumulative sales are also higher. Similarly, a brand with a longer time-to-peak-revenue enjoys a more extended period of growth that contributes to accumulating sales and achieving a higher peak revenue level. Consequently, cumulative sales are again higher (Fischer et al., 2010). The maturity stage is followed by the last stage of "Decline".

Typically, innovative drugs enjoy patent protection for several years after launch. The competitive situation changes dramatically for an innovative drug when its patent expires[3] and cheaper generic competitors rush into the market. In addition, innovative brands may be attacked by other categories offering alternative drug treatment (technological substitution), impacting peak revenues (Fischer et al., 2010). Researchers have studied the impact of patent expiry on drug prices. In one such published study, the researchers indicated that drug prices decreased significantly after patent expiry, with drug price ratios ranging from 6.6 to 66% 1–5 years after patent expiry. The extent of this price reduction varied greatly between products and countries (Vondeling et al., 2018).

Could improvements in product design, development, and manufacturing allow for the breakeven point to be achieved earlier in the product's lifecycle?

[3] The immediate decline in revenue after the patent expiration is referred to as "patent cliff"

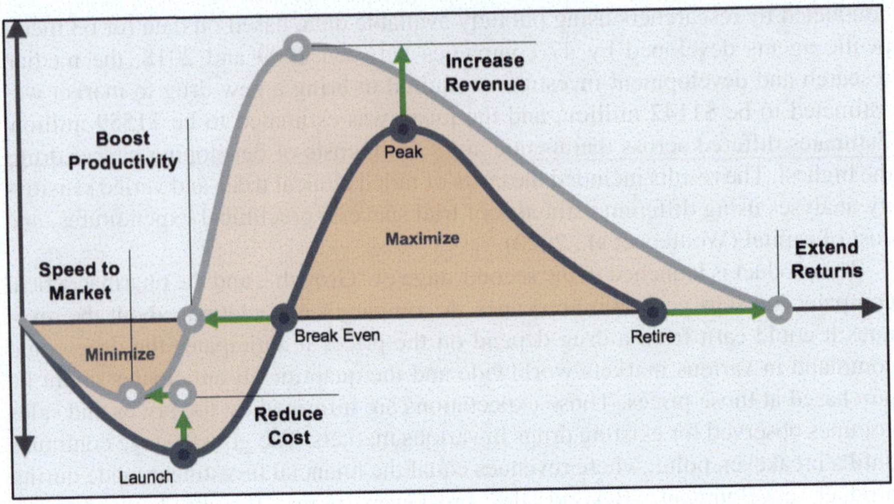

Fig. 5.5 Cost optimized map of cash flow over time and milestones. (CADLOG, 2023)

5.2.2 *Enhancement of Pharmaceutical Product Lifecycle*

The concepts discussed in Fig. 5.4 help in providing the current state of a typical pharmaceutical product's lifecycle. However, there should always be an intent to improve the current state and to enhance the value proposition of pharmaceutical development so that the cost savings can be passed on to the consumers without sacrificing the profitability of the pharmaceutical industry. In such an ideal state, the hypothetical cash flow diagram of Fig. 5.4 could be adapted to a more cost-optimized diagram shown in Fig. 5.5, where areas of potential improvements are identified in each stage of Incubation, Growth, Maturity, and Decline. These potential improvements are discussed in the subsequent sections.

5.3 Opportunities to Reduce Incubation Time

The pharmaceutical industry is very conscious of any development efforts that can help reduce costs and minimize the time it takes for the products to reach the commercial market. Current economic and political pressures are leading many research-based pharmaceutical and biopharmaceutical companies to re-assess decades-old R&D strategies and focus on improving efficiency, boosting output, and meeting unmet patient needs. The changing focus of drug discovery and development and the use of new technologies for identifying and screening candidates for drug development have created new opportunities for the industry to target a host of persistent, degenerative, and life-threatening diseases that have not been targeted in the past.

This changing focus toward targeted medicines, often addressing the needs of relatively small patient populations, is enabled by advances in genomics, proteomics, and other tools that allow researchers to understand disease mechanisms better and identify patient sub-populations.

Potential cost-effectiveness strategies for clinical trials include the following: (1) a focus on core competencies and a higher level of strategic outsourcing, which allowed the company to allocate better and prioritize its resources; (2) active collaboration with global regulatory agencies to ensure that the pre-market studies conducted will meet regulatory demands; (3) enhanced utilization of e-data management technologies, such as electronic data capture, to improve the collection and analysis of clinical data; and (4) increasing conduct of studies in lower-cost and emerging markets, i.e., offshoring (Kaitin, 2010).

However, enhanced drug discovery and efficient clinical trials alone are insufficient to help the industry remain competitive in today's challenging pharmaceutical marketplace. Companies must also address the cumbersome, expensive, and inefficient drug development process. Inefficiencies in drug development can stem from unoptimized drug substance and drug product manufacturing processes leading to low yields and high costs of goods, lengthy stability studies, underestimating the needs of patient compliance, and misaligning the market's needs. The adoption of Quality by Design (QbD) principles where the product is designed to meet patient requirements, the process is evaluated to consistently (and robustly) meet product critical quality attributes, and the impact of starting materials and process parameters on product quality is understood is a good starting point.

5.3.1 Modern Rational Drug Design

Drug design is a complex pharmaceutical science with a long history. Many achievements have been made in drug design since the end of the nineteenth century when Emil Fisher suggested that drug–receptor interaction resembles the key and lock interplay. Drug design has gradually transformed into a coherent, well-organized science with a solid theoretical background and practical applications. Now, drug design is the most advanced approach to drug discovery. It utilizes innovations in science and technology and includes them in its wide-ranging methods and tools to achieve the main goal: discovering effective, specific, non-toxic, safe, and well-tolerated drugs. Drug design is one of the most intensively developing modern sciences, and its progress is accelerated by the implication of artificial intelligence (Doytchinova, 2022).

In the past, serendipity played a significant role in drug discovery. Well-known examples include the discovery of penicillin, chlordiazepoxide (the first benzodiazepine), cyclosporin, and sildenafil (Viagra). The discovery of antibiotics is a fascinating story. The first antibiotic, salvarsan, was deployed in 1910. In just over 100 years, antibiotics have drastically changed modern medicine and extended the average human lifespan by 23 years. The chance discovery of penicillin in 1928[4] started the golden age of natural product antibiotic discovery, leading to a Nobel Prize in Medicine and Physiology (Gaynes, 2017) and peaked in the mid-1950s (Fig. 5.6). Researchers argue that the introduction of antibiotics into clinical use was arguably the greatest medical breakthrough of the twentieth century. In addition to treating infectious diseases, antibiotics have made many modern medical procedures possible, including cancer treatment, organ transplants, and open-heart surgery. However, misuse of these valuable compounds has resulted in the rapid rise of antimicrobial resistance (AMR), with some infections now effectively untreatable (Hutchings et al., 2019).

Another approach for drug discovery is by chemical modifications of known drugs or natural products. Aspirin (acetylsalicylic acid) is an excellent example of this approach. Aspirin is a nonsteroidal anti-inflammatory drug (NSAID) that reduces pain, fever, and inflammation. The discovery of aspirin stretches back more than 3500 years to when bark from the willow tree was used as a pain reliever and antipyretic. Unknown to the ancient Sumerians and Egyptians who used it, salicin was the active agent within willow bark. The active ingredient was discovered in 1828 when Johann Buchner (1783–1852) refined willow bark into yellow crystals. In 1829, the process was further refined by Pierre-Joseph Leroux (1795–1870) in France (Leroux, 1830) and taken a step further in 1838 when Raffaele Piria (1814–1865) produced a stronger compound from the crystals isolated from willow bark, which he named salicylic acid. Acetylsalicylic acid[5] was developed in 1897 by Bayer Company scientists who chemically modified salicylic acid (Fig. 5.7) (Desborough & Keeling, 2017).

Screening of databases, virtually or by high throughput (HTS) assays, is another way to discover new drugs. HTS uses automated equipment to rapidly test thousands to millions of samples for biological activity at the model organism, cellular,

[4] A chance event in a London laboratory in 1928 changed the course of medicine. Alexander Fleming, a bacteriologist at St. Mary's Hospital, had returned from a vacation when he noticed a zone around an invading fungus on an agar plate in which the bacteria did not grow. After isolating the mold and identifying it as belonging to the Penicillium genus, Fleming obtained an extract from the mold, naming its active agent penicillin. He determined that penicillin had an antibacterial effect on staphylococci and other gram-positive pathogens. Fleming published his findings in 1929. At Oxford University, Ernst Chain found Fleming's 1929 article on penicillin and proposed to his supervisor, Howard Florey, that he try to isolate the compound. Penicillin's colossal effects led to the awarding of the Nobel Prize in Medicine and Physiology in 1945 to Fleming, Chain, and Florey. Penicillin was isolated from other microorganisms, which led to a new term, antibiotics. Using similar discovery and production techniques, researchers discovered many other antibiotics in the 1940s and 1950s: streptomycin, chloramphenicol, erythromycin, vancomycin, and others (Gaynes, 2017).

[5] Acetylsalicylic acid is a prodrug and the more tolerated acetyl derivative of salicylic acid. It took until the early 1970s to identify the inhibition of prostaglandin biosynthesis as the primary mode of action of acetylsalicylic acid.7 Sune K. Bergström, Bengt I. Samuelsson, and John R. Vane were awarded the Nobel Prize in Physiology or Medicine in 1982 for this discovery, which also led the way for the use of low-dose Aspirin as an antiplatelet agent for the prevention of cardiovascular diseases, such as coronary artery disease, heart attack, and stroke.

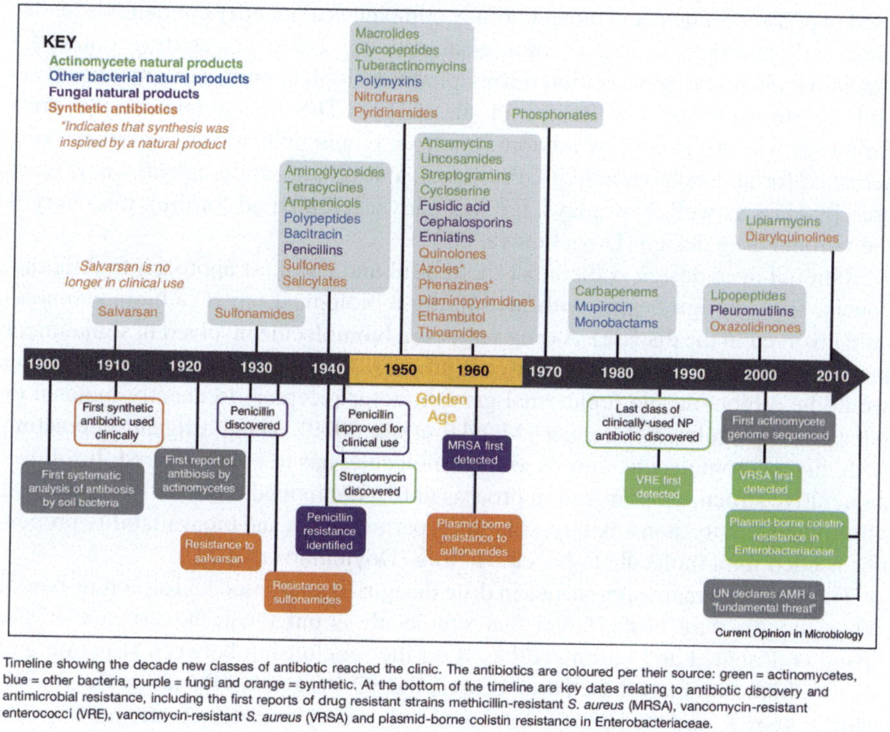

Fig. 5.6 Timeline showing the introduction of new classes of antibiotics into clinic. (Hutchings et al., 2019)

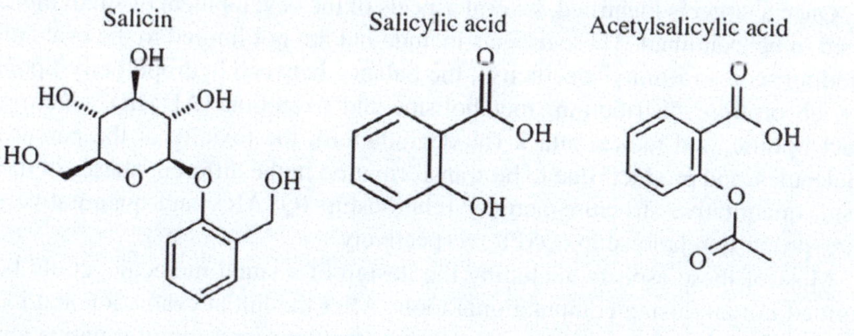

Fig. 5.7 Molecular structure of salicin, salicylic acid, and acetylsalicylic acid. (Desborough & Keeling, 2017) (Reprinted with permission from John Wiley and Sons)

pathway, or molecular level. In its most common form, HTS is an experimental process in which 10^3–10^6 small molecule compounds of known structure are screened in parallel. Other substances, such as chemical mixtures, natural product extracts, oligonucleotides, and antibodies, may also be screened. HTS is commonly used in pharmaceutical and biotechnology companies to identify compounds (called hits) with pharmacological or biological activity. These are starting points for medicinal chemical optimization during pharmacological probes or drug discovery and development (Attene-Ramos et al., 2014). The first sulphonamide drug, Prontosil, was discovered by random in vitro screening when many colorants were screened for antibacterial activity. Paclitaxel, a novel anti-tumor agent, was discovered by HTS as well. Nowadays, the most advanced method for drug discovery is the rational drug design (Doytchinova, 2022).

Rational drug design is the most thoughtful and cheapest approach to drug discovery. Drug design begins with identifying a biological target (a biomacromolecule involved in the disease). A drug target is a biomolecule involved in signaling or metabolic pathways specific to a disease process. As a prime example, a drug target would be a biomolecule (epidermal growth factor receptor) frequently mutated or otherwise deregulated in cancer (Mandal et al., 2009). Then, a ligand interacting with this macromolecule, known as a hit molecule, has to be discovered. It follows an iterative structure optimization process until a compound is derived with optimal affinity, selectivity, non-toxicity, solubility, permeability, and bioavailability properties needed for a molecule to become a drug (Doytchinova, 2022).

There are two main approaches in drug design: ligand-based and structure-based. When the structure of the target macromolecule is unknown, the structure of the ligand is designed and optimized based on the relationship between structure and activity. In the structure-based drug design, the 3D structure of the target macromolecule is known, and the ligand is designed to be complementary to the binding site on the macromolecule. Complementarity means a steric, electrostatic, and hydrophobic fit between the ligand and the target. Steps related to these two approaches and evaluating other properties in rational drug design are presented in the following flow charts (Fig. 5.8) (Mandal et al., 2009).

Once a target is identified, several aspects of the development of small molecules need to be examined. These aspects include but are not limited to the evaluation of binding scores (affinity/specificity), the balance between hydrophilicity/lipophilicity, absorption, distribution, metabolism, and excretion (ADME), electrophilic, nucleophilic, and radical attack (biodegradation), the toxicity of the parent small molecules, and products due to biotransformation in the different phases of metabolism, quantitative structure–activity relationship (QSAR), and quantitative structure–property relationship (QSPR) respectively.

Most of these aspects, including the design of a small molecule, could be performed initially using computational tools. After the initial evaluation and identification of lead molecules, gene expression profiling, and bioinformatics analysis would be particularly important to gain insights into gene expression patterns. In turn, this knowledge can be utilized to improve drugs to accomplish desirable attributes such as disease-free survival, eradication of disease, elimination or minimization of toxic side effects, reduction of undesirable biotransformation, improvement in distribution (bioavailability), overcoming of drug resistance, and improvement of immune responses. Therefore, rational drug design is integral to drug development and discovery (Mandal et al., 2009).

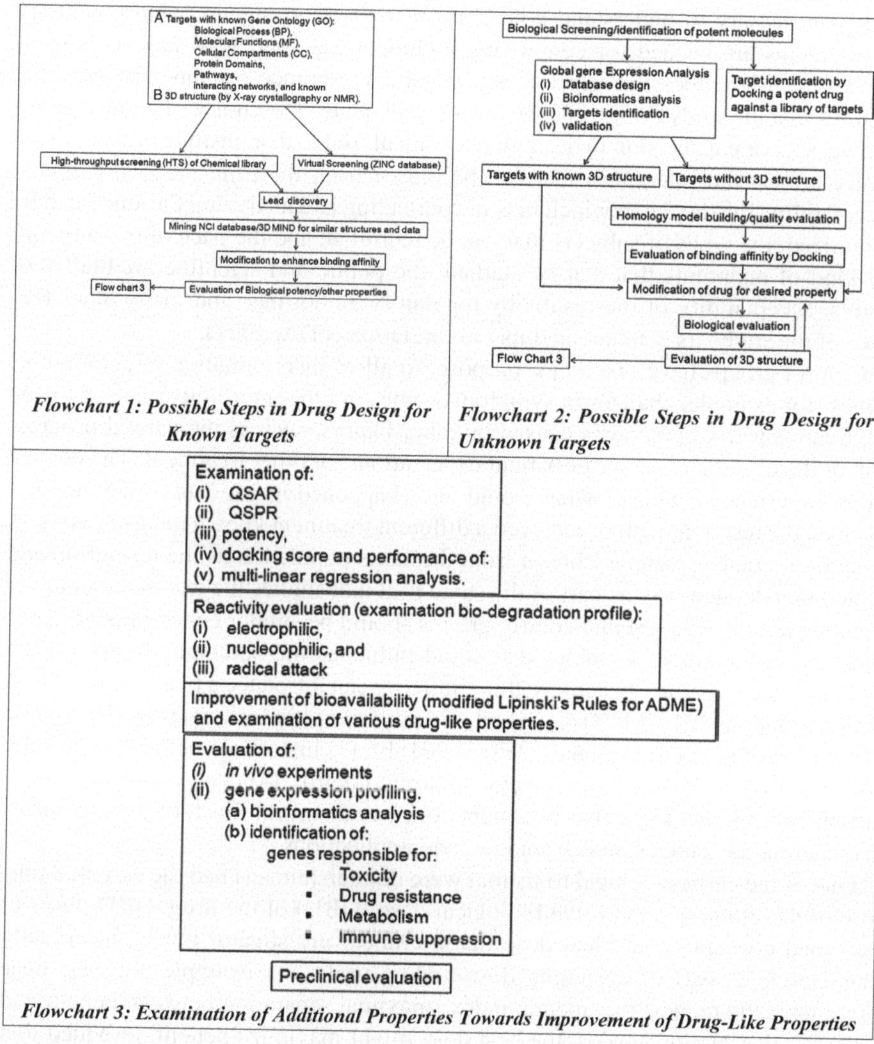

Fig. 5.8 Flowcharts followed in drug design. (Mandal et al., 2009) (Reprinted with permission from Elsevier)

5.3.2 *Improvements in Clinical Trial Designs*

Before a new drug or biologic can be marketed, its sponsor must show, through adequate and well-controlled clinical studies, that it is effective. A well-controlled study permits a comparison of subjects treated with the new agent with a suitable control population so that the new agent's effect can be determined and distinguished from other influences, such as spontaneous change, "placebo" effects, concomitant therapy, or observer expectations. Depending on the stage of product development and statistical design of the study, the cost of clinical trials can range from a few million dollars to even hundreds of millions of dollars.

It is necessary to understand how clinical trials are designed, what regulatory requirements are needed for conducting a clinical trial, and how success is measured. The FDA does not express any general preference for any one type but requires that the study design chosen must be adequate. The choice of control group is always a critical decision in designing a clinical trial to demonstrate efficacy. That choice affects the inferences that can be drawn from the trial, the trial's ethical acceptability, the degree to which bias in conducting and analyzing the study can be minimized, the types of subjects that can be recruited, and the pace of recruitment, the kind of endpoints that can be studied, the public and scientific credibility of results, acceptability of the results by regulatory authorities, and many other features of the study, its conduct, and its interpretation (FDA, 2001).

Control groups have one major purpose: to allow discrimination of patient outcomes (for example, changes in symptoms, signs, or other morbidity) caused by the test treatment from outcomes caused by other factors, such as the natural progression of the disease, observer or patient expectations, or other treatment. The control group experience evaluates what would have happened to patients if they had not received the test treatment or received a different treatment known to be effective. A concurrent control group is chosen from the same population as the test group and treated in a defined way as part of the same trial that studies the test treatment over the same period. The test and control groups should be similar concerning all baselines and on-treatment variables that could influence the outcome, except for the study treatment. Failure to achieve this similarity can introduce a bias into the study. FDA regulations [21 CFR 314.126] cite five different kinds of controls that can be useful in particular circumstances (FDA, 2018b): (1) historical control, (2) placebo concurrent control, (3) no-treatment concurrent control, (4) dose-comparison concurrent control, and (5) active-treatment concurrent control. Each of these controls has inherent advantages, disadvantages, and applications.

One of the earliest clinical trials that were done in humans had the essential purpose of determining an optimal biological dose (OBD) of the drug. OBD refers to the broad concept of an ideal dose that a clinical investigator must conceptually characterize as part of designing dose-finding trials. For example, the best dose might have the highest therapeutic index (maximal separation between benefit and risk). In other circumstances, the best dose might maximize benefit, provided that risk is below some pre-specified threshold. The OBD also depends on the clinical circumstances and purposes of treatment. For example, the optimal dose of a new analgesic might be the lowest dose that completely relieves mild to moderate pain in 90% of recipients. This is a type of minimum effective dose (MED).

In contrast, the optimal dose of a new antibiotic to treat severe infections and discourage resistance might be the highest dose that causes significant side effects in less than 5% of patients. This is a type of maximum non-toxic dose (MND). For cytotoxic drugs to shrink tumors, it has been thought historically that more is better, leading investigators to push doses as high as can be tolerated. Such a dose that yields serious (even life-threatening) but reversible toxicity in no more than 30% of patients is a maximum tolerated dose (MTD). The optimal dose for developing a molecular targeted agent might be the dose that suppresses 99% of the target

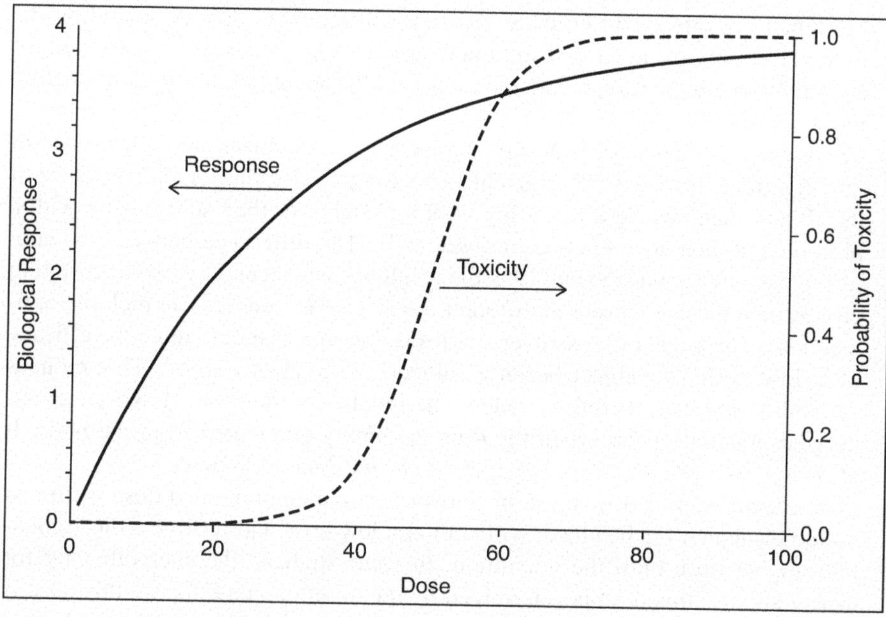

Fig. 5.9 Relationship of potential toxicity and biological response to dose. (Piantadosi, 2017) (reprinted with permission from John Wiley and Sons)

activity in at least 90% of patients. Until definitive clinical data are obtained, this might be viewed as a most likely-to-succeed dose (MLS) (Piantadosi, 2017).

Characterizing the OBD is a critical conceptual step that will suggest specific experimental design features for dose-finding. Dose-finding (DF) studies are a type of treatment mechanism trial and focus on the OBD or schedule of drug, estimating pharmacokinetic parameters that relate to dose and safety and the frequency of side effects. DF studies are conducted with sequentially rising doses in successive cohorts and terminate when specific predefined clinical outcomes are observed. A potentially more difficult problem arises when balancing the toxicity of a new drug against its biological marker response or intermediate outcome. This situation is shown hypothetically in Fig. 5.9, where both the probability of toxicity (right-hand axis) and biological response (left-hand axis) are plotted against the dose. The two effect curves are not directly comparable because the respective vertical axes are different. An optimum dose would produce low toxicity while nearly maximizing the biological response. Such a dose is not guaranteed to exist (Piantadosi, 2017).

The four most common types of clinical study designs are discussed below (White, 2013).

- ***Parallel Dose Comparison:*** Parallel dose comparison studies are classical dose-finding studies and are still one of the most common study designs. In a parallel dose escalation study, several potential doses are selected, and subjects are randomized to receive one of the doses or placebos for the entire study. At the end of the study, each treatment group can be compared to the control group and examined for safety and efficacy. Because all treatment groups, including the higher dose cohorts, are dosed simultaneously, this study design is best suited for situations where the researchers have a good idea about the safety profile before the study starts.

- ***Cross-over:*** In a cross-over design, subjects are randomized into a sequence of investigational products (IP) and a placebo. Specifically, they are given a dose of the IP and then switched to dosing with a placebo, or they start dosing with a placebo and then are switched to doses of IP. The difference between the subjects' response to placebo and IP is the result of interest, and by having different groups of subjects exposed to different doses, the researchers can pick the optimal dose. The value of cross-over studies is that they can determine the efficacy of a dose within a subject because subjects act as their control. This reduces variability and can, therefore, reduce the number of subjects. However, cross-over designs only work when the drug is quickly eliminated from the body. It also requires a product that is designed to be used multiple times.
- ***Dose Titration:*** In a dose titration study, the maximum tolerated dose is titrated within a subject. Each subject will start at a low dose and receive an incrementally higher dose until the maximum. In some studies, like chemotherapy for cancer studies, this dose is determined by the onset of side effects—this dose is called the MTD. In other studies where the product is less toxic, it may depend on the blood levels of the IP, a metabolite, or a maximum dose determined from preclinical studies. Dose titration studies work well for treatments of chronic conditions where a drug will be used for an extended period and where the dose is likely to be tailored to the subject's weight or reaction. This design is also suitable for situations where it is expected that significant differences will be observed in the way each subject reacts. Chronic hypertension medications are an excellent example of products where dose titration is useful.
- ***Dose Escalation:*** If researchers are unsure of a drug's safety profile and want to start exposing subjects to lower doses first, a dose escalation study is a good design. In this study, researchers begin with one group of subjects (often referred to as a cohort) and give them a low dose. This group is observed for some time, and if no safety issues are noted, a new group of subjects is enrolled and given a higher dose. This process is repeated until either the researchers reach the MTD or they reach the highest dose is planned. This design increases patient safety because it can start by exposing a small number of subjects to the lowest dose possible.

In pharmaceutical/clinical research and development, clinical trials evaluate the safety and efficacy of a test treatment under investigation. In the past several decades, it has become evident that increased spending does not reflect an increased success rate in new product development. As a result, the FDA kicked off a Critical Path Initiative in the early 2000s to assist sponsors in identifying the scientific challenges underlying the medical product pipeline problems. In 2006, the FDA released a Critical Path Opportunities List that outlines 76 initial projects (six broad topic areas) to bridge the gap between the quick pace of new biomedical discoveries and the slower pace at which those discoveries are currently developed into therapies. Among the 76 projects, the FDA called for advancing innovative trial designs, especially for using prior experience or accumulated information in trial design, which was interpreted as an encouragement for using adaptive design methods in clinical trials (Chow & Corey, 2011).

Traditionally, clinical trials have been run in three steps: (1) the trial is designed, (2) the trial is conducted as prescribed by design, and (3) once the data are ready, they are analyzed according to a pre-specified analysis plan. This practice is straightforward but inflexible as it does not include options for changes that may become desirable or necessary during the trial. Adaptive designs (ADs) provide an alternative (Pallmann et al., 2018). In 2019, the FDA issued new guidance to sponsors and applicants submitting INDs, NDAs, BLAs, or supplemental applications on the appropriate use of adaptive designs for clinical trials to provide evidence of the effectiveness and safety of a drug or biologic (FDA, 2019).

Adaptive clinical trials add a review-adapt loop to the linear design-conduct-analysis sequence, as shown in Fig. 5.10. Scheduled interim looks at the data are allowed while the trial is ongoing, and pre-specified changes to the trial's course can be made based on analyses of accumulating data while maintaining the validity and integrity of the trial. Pre-planned changes that an AD may permit include, but are not limited to:

- refining the sample size
- abandoning treatments or doses
- changing the allocation ratio of patients to trial arms.

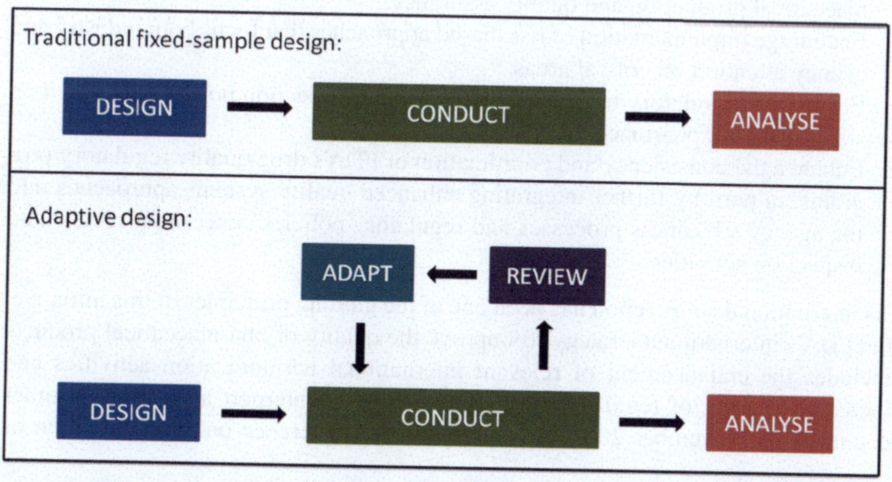

Fig. 5.10 Traditional clinical trial design vs. adaptive clinical design. (Pallmann et al., 2018)

- identifying patients most likely to benefit and focusing recruitment efforts on them.
- stopping the whole trial at an early stage for success or lack of efficacy.

Therefore, adaptive designs can make clinical trials more flexible by utilizing results accumulating in the trial to modify the trial's course per pre-specified rules. Trials with an adaptive design are often more efficient, informative, and ethical than trials with a traditional fixed design since they often make better use of resources such as time and money and might require fewer participants. Adaptive designs can be applied across all phases of clinical research, from early-phase dose escalation to confirmatory trials (Pallmann et al., 2018).

5.3.3 Adoption of QbD

In August 2002, the FDA announced a significant new initiative, Pharmaceutical Current Good Manufacturing Practices (cGMPs) for the twenty-first century, to enhance and modernize the regulation of pharmaceutical manufacturing and product quality—to bring a twenty-first-century focus to this critical FDA responsibility. The initiative was intended to modernize the FDA's regulation of pharmaceutical quality for veterinary and human drugs and select human biological products such as vaccines. As part of this initiative, pharmaceuticals and CMC regulatory programs were evaluated with the following objectives.

- Encourage the early adoption of new technological advances by the pharmaceutical industry.
- Facilitate industry application of modern quality management techniques, including implementation of quality systems approaches, to all aspects of pharmaceutical production and quality assurance.
- Encourage implementation of risk-based approaches that focus both industry and agency attention on critical areas.
- Ensure that regulatory review, compliance, and inspection policies are based on state-of-the-art pharmaceutical science.
- Enhance the consistency and coordination of FDA's drug quality regulatory programs, in part, by further integrating enhanced quality systems approaches into the agency's business processes and regulatory policies concerning review and inspection activities.

As international cooperation has been one of the guiding principles of this initiative, the FDA's international strategy to improve the quality of pharmaceutical products includes the enhancement of relevant international harmonization activities and increased sharing of regulatory information with counterpart authorities in other countries. In November 2003, the International Conference on Harmonization of

Table 5.2 Main ICH quality guidelines defining the QbD framework (Destro & Barolo, 2022) (Reprinted with permission from Elsevier)

Year	Guideline
2000	Q7—Good manufacturing practice guide for active pharmaceutical ingredients
2005	Q9—Quality risk management system
2008	Q10—Pharmaceutical quality system
2009	Q8(R2)—Pharmaceutical development
2010	Q8, Q9, Q10 Questions and Answers
2011	Q8, Q9, Q10 Points to consider
2012	Q11—Development and manufacture of drug substances (chemical entities and biotechnological/biological entities)
2017	Q11 Questions and answers
2018	Q14—Analytical procedure development
2019	Q12—Technical and regulatory considerations for pharmaceutical product lifecycle management
2021	Q13—Continuous manufacturing of drug substances and drug products

the Technical Requirements for Registration of Pharmaceuticals (ICH)[6] agreed to work on a harmonized plan to develop a pharmaceutical quality system based on an integrated approach to risk management and science (FDA, 2004). The result was that pharmaceutical QbD has evolved with the issuance of ICH Q8 (R2) (Pharmaceutical Development), ICH Q9 (Quality Risk Management or QRM), ICHQ10 (Pharmaceutical Quality System or PQS), and ICH Q11 (Development and Manufacture of Drug Substance) have been issued. These documents provide high-level directions concerning the scope and applications of QbD to the pharmaceutical industry. An informative summary of these documents was compiled by (Destro & Barolo, 2022) and is presented in Table 5.2.

Pharmaceutical QbD is a systematic approach to development that begins with predefined objectives and emphasizes product and process understanding and control based on sound science and quality risk management. The goals of pharmaceutical QbD may include the following:

- To achieve meaningful product quality specifications that are based on clinical performance.
- To increase process capability and reduce product variability and defects by enhancing product and process design, understanding, and control.

[6] Harmonization of regulatory requirements was pioneered by the EC, Europe, in 1980s, as EC, Europe moved towards developing a single market for pharmaceuticals. The success achieved in Europe demonstrated that harmonization was feasible. At the same time there were discussions between Europe, Japan and USA on possibilities for harmonization. It was at the WHO Conference of Drug Regulatory Authorities (ICDRA), in Paris, in 1989, that specific plans for action began to materialize. Soon afterwards, the authorities approached International Federation of Pharmaceutical Manufacturers and Associations (IFPMA) to discuss a joint regulatory-industry initiative on international harmonization, and ICH was conceived.

- To increase product development and manufacturing efficiencies
- To enhance root cause analysis and post-approval change management

Under QbD, these goals can often be achieved by linking product quality to the desired clinical performance and then designing a robust formulation and manufacturing process to consistently deliver the desired product quality (Yu et al., 2014). Table 5.3 illustrates some potential contrasts between what might be considered a minimal approach and an enhanced QbD approach regarding different aspects of pharmaceutical development and life cycle management. Current practices in the pharmaceutical industry vary and typically lie between the two approaches presented in Table 5.3. There is no doubt that the introduction of QbD has been a

Table 5.3 Minimal vs. QbD approach to product development (ICH, 2009)

Aspect	Minimal approach	QbD approach
Overall pharmaceutical development	Mainly empirical developmental research is often conducted one variable at a time.	Systematic, relating mechanistic understanding of material attributes and process parameters to drug product critical quality attributes Multivariate experiments to understand product and process Establishment of design space Process analytical technology (PAT) tools utilized
Manufacturing process	Fixed Validation is primarily based on initial full-scale batches Focus on optimization and reproducibility	Adjustable within the design space Life cycle approach to validation and, ideally, continuous process verification Focus on control strategy and robustness Use of statistical process control methods
Process controls	In-process tests primarily for go/no-go decisions Off-line analysis	PAT tools utilized with appropriate feed-forward and feedback controls Process operations tracked and trended to support continual improvement efforts postapproval
Product specifications	Primary means of control Based on batch data available at the time of registration	Part of the overall quality control strategy Based on desired product performance with relevant supportive data
Control strategy	Drug product quality is controlled primarily by intermediates (in-process materials) and end-product testing	Drug product quality is ensured by a risk-based control strategy for well-understood products and processes Quality controls shifted upstream, with the possibility of real-time release testing or reduced end-product testing
Life cycle management	Reactive (i.e., problem-solving and corrective action)	Preventive action Continual improvement facilitated

welcome evolution in the field of pharmaceutical product development and manufacturing. The industry has enthusiastically embraced key concepts of QbD, and case studies involving the integration of QbD in product development are regularly published in pharmaceutical literature.

5.3.4 *Patient-Centric Dosage Design and Improved Medication Adherence*

Patient-centric pharmaceutical drug product design describes an approach that directly aligns product characteristics with patient characteristics for a therapeutic goal in a targeted patient population(s) and requires an appropriate definition that can be used consistently among key stakeholders. It can be defined as identifying the comprehensive needs of individuals or the target patient population and utilizing the identified needs to design pharmaceutical drug products that provide the best overall benefit to risk profile for that target patient population over the intended duration of treatment. The definition also addresses the requirement that human (patient) characteristics are considered in the product design. In this sense, for example, a primary package closure system should not only fulfill its functional requirement of protecting the product against environmental contamination and ensuring product stability but also incorporate requirements such as being easy to open and reclose and facilitating accurate dose measurement by caregivers and patients with limited manual dexterity, grip strength, or visual capacity (Stegemann et al., 2016).

Based on the definitions of "pharmaceutical drug product" and "patient centricity," a patient-centric pharmaceutical drug product design spans from the basic concept of a pharmaceutical drug product to the its performance and effectiveness in the patient's treatment. This drug product-patient interface includes elements that lead to the intended (re)action of the patients to use the product as intended and prevent medication non-adherence (intentional) and medication errors (unintentional). Patient-centric pharmaceutical drug product design, therefore, cannot solely be based on scientific or technical theories but will have to acknowledge the layperson's response to the product design. Therefore, testing the drug product design with the targeted patient population within their personal health and environmental context will be essential to developing high-quality, patient-centric drug products. In an excellent manuscript by (Stegemann et al., 2016), a thorough analysis was performed to detail the various aspects of patient-centric pharmaceutical drug product design. The authors summarized the primary drug product-related and patient-related characteristics for consideration in a patient-centric pharmaceutical drug product design and development process (Table 5.4). Numerous product-related and patient-related characteristics can help develop a patient-centric pharmaceutical drug product.

A significant advantage of patient-centric dosage form design is the improvement of medication adherence. Patients are considered adherent to medications when they take prescribed agents at doses and times recommended by a healthcare provider and agreed to by the patient. Medication non-adherence is a significant

Table 5.4 Characteristics for consideration of patient-centric pharmaceutical drug product design (Stegemann et al., 2016)

Product-related characteristics	Patient-related characteristics
• Type of drug substance (drug profile, desired effects, side effects, therapeutic window, mechanism of action) • Dose to therapeutic effect (required dose range for different patient populations; need for small or larger dosing increments, dose flexibility) • Route of administration (oral, nasal, rectal, parenteral, …) • Type of dosage form (tablet, capsule, solution, injection, …) • Formulation characteristics (immediate release, modified release, tablet size/shape/color./strength/concentration/, excipient composition, …) • Product strength/concentration • Packaging (primary, secondary, tertiary) • Dosing and administration devices (oral syringe, inhalator, …) • Product recognition, identification (appearance unpacked product and in its primary, secondary, or tertiary packaging name, name…) • Instructions for use (SmPC, PIL, labeling, others….) O Specific dosing requirements (e.g., before meal, not together with drug *x*, …) O Product stability O Storage conditions O In-use shelf life and storage conditions O Preparation steps to administration O Compatibility with diluents and other medicines where appropriate O Compatibility with food or drink O Industry-verified manipulations of the dosage form	• Developmental stage/age O Newborn, toddler, … O Organ and body functions (kidney, liver, lung …) O Social-emotional development O Way of living (alone, with spouse, with parents) O Access to caregivers • PK/PD and physiological changes (e.g., receptor density, kidney clearance, …) • Visual impairment O Near acuity O Visual field O Blindness • Loss of hearing • Swallowing impairment O Safe solid dose swallowing O Dysphagia • Poor hand sensitivity O Control of movement O Control of strength O Sensing surfaces or temperature O Control of eye-hand coordination • Motoric impairment O Grip strength O Pinch strength O Manual dexterity O Arm mobility (e.g., lift above head) O Difficulties walking O Bedridden • Cognitive impairment O Mild dementia, memory loss O Information processing speed (hearing, understanding) • Health literacy • Dentition • Psychological issues O Adherence O Negative perception O Depressive disorders • Disease state, co-morbidity O Experience of disease O Disease cluster O Disabilities O End point of disease progression

public health consideration, leading to poor health outcomes, increased healthcare service utilization, and healthcare costs. Researchers claim that 20–30% of dollars spent in the US healthcare system are wasteful (Iuga & McGuire, 2014). Research has shown that poor medication adherence is widespread and varied by disease, patient characteristics, and insurance coverage. Studies have consistently shown that 20–30% of medication prescriptions are never filled and that approximately 50% of medications for chronic disease are not taken as prescribed. This lack of adherence dramatically affects health outcomes and costs (Fig. 5.11). In the United States, it is estimated to cause approximately 125,000 deaths, at least 10% of hospitalizations, and a substantial increase in morbidity and mortality per year.

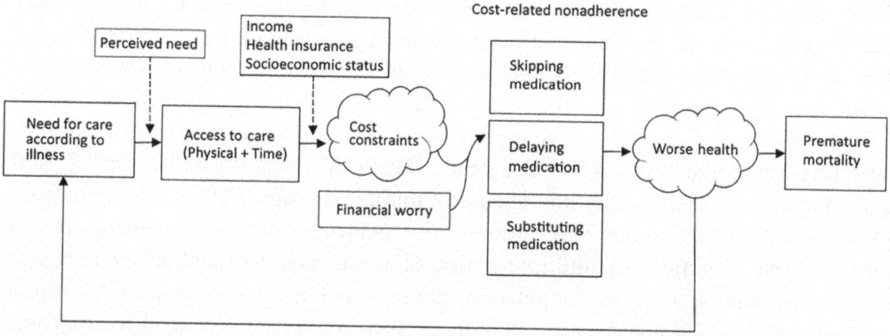

Fig. 5.11 Conceptual model of the relationship between cost-related non-adherence and mortality.) (Van Alsten & Harris, 2020)

Non-adherence has been estimated to cost the US healthcare system between $100 billion and $289 billion annually (Viswanathan et al., 2012).

Literature provides multiple definitions for patient compliance. In one definition, compliance is the 'extent to which a person's behavior (in terms of taking medications, following diets, or executing lifestyle changes) coincides with medical or health advice.' Thus, by this definition, compliance is viewed as a process. In another interpretation, compliance is viewed in terms of medication results. Compliance as an outcome is defined as 'the number of doses not taken or taken incorrectly that jeopardize the therapeutic outcome' or 'the point below which the desired preventive or desired therapeutic result is unlikely to be achieved. Outcome-oriented definitions differ from the process-oriented definition in emphasizing the result or outcome of the actions taken (Morris & Schulz, 1992). From the pharmaceutical dosage form design viewpoint, an outcome-orientated definition is more applicable. There are numerous reasons for non-compliance from a patient when dealing with an outcome-oriented approach for a solid oral dosage form, ranging from the length of therapy, dosage form size, product packaging, and the similarity of different products. No matter the reason for a patient's poor (or complete lack of) compliance, the costs incurred are huge! However, pharmaceutical formulation scientists can aid in improving healthcare outcomes by incorporating aspects of patient-centricity in the design of their products. One such way this can be further accomplished is by the development of fixed-dose combination (FDC) products that reduce the instances of polypharmacy (use of ≥5 prescription drugs).

In the medical setting, combination therapy is an important treatment modality for many diseases, including cancer, cardiovascular disease, and infectious diseases. Recent scientific advances have increased the medical communities' awareness of the pathophysiological processes that underlie these and other complex diseases.

This increased understanding has provided further impetus to develop new therapeutic approaches using combinations of drugs directed at multiple therapeutic targets to improve treatment response, minimize the development of resistance, or minimize adverse events. In situations where combination therapy provides significant therapeutic advantages, there is growing interest in developing combinations of new investigational drugs. Over the years, the FDA has published essential guidance documents on this topic. In the case of marketed drugs, FDCs[7] offer benefits to pharmaceutical companies and patients. For pharmaceutical companies, creative and effective matching of multiple active pharmaceutical ingredients (APIs) can open new markets, whereas, for patients, FDCs can offer convenience and therapeutic benefits. From a healthcare outcome perspective, FDCs can lead to improved pharmacotherapies, an extended range of treatment options, synergy, elongated product life cycle management, cost savings, and patient compliance. Often, FDCs are composed of previously approved agents, and the 505(b)(2) pathway is commonly used for these approvals (King, 2017). The approval rate by the FDA for FDCs is steadily going up and averages around 10% of the total NDA approvals between 2010 and 2015. Most FDC approvals were a dual (and triple) combination of approved molecules, amounting to almost 55% of all FDC approved (Kwon & Lee, 2017).

5.3.5 *Adaptation of Digital Strategies and Tools*

The term Pharma 4.0 has been introduced recently, referring to adapting the digital strategies and tools of Industry 4.0 principles and their application to pharmaceutical manufacturing and supply chain practices. In this context, digital tools and orchestration platforms are being developed under Industry 4.0/5.0 principles. The term refers to manufacturing digitalization and automation of processes, introducing autonomous, computerized systems. It utilizes mathematical models (e.g., statistical, kinetic) and the Internet of Things to facilitate and maintain internal communication within and across factories. Application of Industry 4.0/5.0

[7] From a drug product perspective, an FDC includes two or more drug substances (whether approved or investigational) combined in a single dosage form, which is manufactured and distributed in fixed doses. Terms like "combination drug" or "combination drug product" can be common shorthand for an FDC product (because most combination drug products are currently FDCs), although the latter is more precise if in fact referring to a mass-produced product having a predetermined combination of drugs and respective dosages (as opposed to customized polypharmacy via compounding). And it should also be distinguished from the term "combination product" in medical contexts, which without further specification can refer to products that combine different types of medical products—such as device/drug combinations as opposed to drug/drug combinations.

principles aims to facilitate: (a) data collection, analysis, and interpretation, (b) man-machine cooperation, (c) online monitoring and control, and (d) intra- and inter-facility data sharing. In the last few years, we have seen the emergence of cloud-based applications to assist decision-making in the pharmaceutical industry. Several industrial players have embraced Pharma 4.0 either through the development of digital platforms to be used by manufacturers or by integrating digitalization into their manufacturing processes (Sarkis et al., 2021).

5.4 Opportunities to Enhance Growth by Boosting Productivity

When an innovative pharmaceutical company launches a new product, it is a cause for celebration as the new drug could help patients suffering from rare disorders to common diseases—to gain new hope for improved quality of life and, in some cases, improved chances of surviving life-threatening illnesses. However, achieving a successful launch is not always straightforward and requires diligent planning. In the US, after the drug company submits the NDA or BLA, the FDA performs its scientific analyses. As part of this review, the FDA must examine the drug's professional labeling and inspect the facility where the drug is manufactured. An FDA approval at this stage means that the drug has met the necessary safety and efficacy data, labeling, and manufacturing standards. Once the FDA confirms that the deemed safety and efficacy standards have been acceptably met, it can authorize market approval nationally. After this step, drug companies must negotiate pricing with private insurance companies or Medicaid/Medicare to generate revenue (Thind & Kowey, 2020). Once the decision on the reimbursement rate is made, the demand for new drugs might still be influenced in several ways, such as education, supply-chain engineering, economic evaluation, and enforcement (Lublóy, 2014).

One of the most essential considerations when first introducing the product into the market is the time-to-market (TTM) metric. Once a drug is approved, it can be marketed under the protection of the patent. Since patents are filed early in the R&D phase when promising compounds are found, only a short time remains before the patent expires. As soon as the drug goes off patent, generic manufacturers launch similar drugs, which drive the drug's price down. Hence, developers of new pharmaceutical drugs strongly focus on reducing TTM, which prolongs the profitable period under patent protection. The importance of TTM is illustrated in Fig. 5.12. Assuming that the market penetration curve is the same regardless of the launch date, a launch at T1 increases the time under patent protection compared to a launch in T2, and the company thereby extends the peak revenue. In practice, the peak

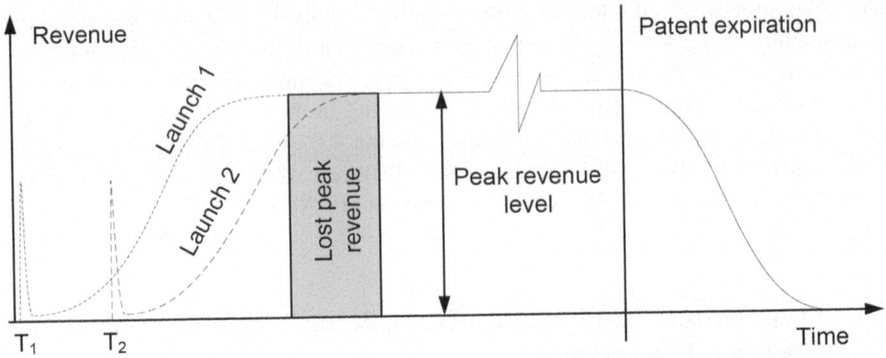

Fig. 5.12 Illustration of lost peak revenue due to delayed commercial launch. (Hansen, 2014)

revenue level may also be lower for a launch in T2 as the market share may be reduced as competitor products possibly have gained ground during the delay (Hansen, 2014). In a study conducted by Deloitte, researchers analyzed the actual and forecast sales for novel drugs approved in the United States between 2012 and 2017 and found a wide variability in launch performance. As per the study, around 36% of all new drugs launched failed to meet market expectations in their first year. At the same time, around 26% of new drug launches far exceeded expectations. These researchers highlighted numerous factors for poor launch performance (Ford et al., 2020b). Their findings are summarized in Table 5.5, and a few factors that impact product design, development, and manufacturing are discussed in detail in the following subsections.

5.4.1 *Increase Access to Global Markets*

In healthcare, market access has been defined as the set of strategies, activities, and processes that pharmaceutical and biomedical companies develop to ensure that their products are made available and adequately priced in a specific healthcare system. Essentially, market access is synonymous with patient access, ensuring that patients have the right access to the right therapy at the right price and the right time. The role of a manufacturer's market access team is to secure appropriate payer coverage for their products. The teams must also lay the groundwork for each drug's pricing, contracting, reimbursement, and fulfillment strategies. The cornerstone of a manufacturer's market access strategy is defining and communicating a product's clinical and economic value to payers, providers, and patients, as each is involved in the decision to initiate therapy (Zele, 2023).

Market access is a critical function of a drug's development and go-to-market strategy, as it sets the stage for a successful product launch. For many

Table 5.5 Factors impacting new product launch performance

Factors	Details
Limited market access	Unfavorable placement or exclusion from formulary Weak health-economic evidence to support price. Higher than expected expenses on discounts and rebates
Inadequate understanding of market and customer needs	Lack of understanding of the types of patients who would respond well to the treatment. Lack of understanding of the types of physicians who would be high versus low prescribers for the new drug. Underestimation of the difficulty of converting customers (physicians and patients) from existing therapies, particularly in competitive drug classes or when generic treatments are available
Poor product differentiation	The product value proposition was poor. Product formulation (inconvenient dosing, need for titration, or in-office administration) created hurdles in patient compliance.
Unfavorable safety profile	Black box warning Physicians' reluctance to prescribe due to an unexpected finding in a clinical trial
Low prioritization or insufficient resource allocation	Inadequate sales force deployment There are no support programs in place to help early adopters with prior authorization hurdles. Leaving out large segments of prescribers from awareness and education campaigns until later
Unexpected events	Patent litigation Manufacturing delays Supply-chain issues

Adapted from (Ford et al., 2020b)

pharmaceutical companies, planning for commercialization only begins in earnest when a drug has been submitted for FDA approval—which is far too late. Ideally, a company's commercial and market access teams should be fully functional long before approval, as many decisions must be made in the early phases of the drug development life cycle. For a successful launch, manufacturers must start market access planning 12–18 months before FDA approval. Otherwise, companies risk launching their product without firmly establishing their coverage, distribution, or reimbursement strategies—which can lead to low uptake and suboptimal sales figures upon launch (Zele, 2023).

In terms of current trends for sales, a report (that included data from IQVIA) compiled by the European Federation of Pharmaceutical Industries and Associations (EFPIA)[8] states that 64.4% of sales of new medicines launched during the period 2016–2021 were on the US market, compared with 16.8% on the European market

[8] EFPIA represents the research-based pharmaceutical industry operating in Europe. Founded in 1978, its members comprise 37 national pharmaceutical industry associations, 38 leading pharmaceutical companies and 17 small and medium sized enterprises undertaking research, development and manufacturing of medicinal products in Europe for human use.

Fig. 5.13 Geographical breakdown of sales of new medicines launched during 2016–2021. (EFPIA, 2022)

(top 5 markets[9]). The numbers were significantly lower for the rest of the world, as shown in Fig. 5.13. The opportunity to generate sales globally is also hampered by launch delays[10] of new drugs in global markets. In a study of the impact of launch delay of new drugs in China, researchers found that Korea and Japan had median delays of 28.2 months and 54.1 months, respectively, between 2009 and 2017. By contrast, for the same timeframe, the median delay in China was 48.70 months during 2009–2017, with the longest delay exceeding 16 years (Zhu & Liu, 2020). Launch delays may be unavoidable due to varying complex factors such as business strategy, concerns around intellectual property rights, price regulations, and country-specific laws. However, any proactive efforts that pharmaceutical companies can take to reduce launch delays could help increase the percentage of global sales while also providing significant benefits to patients by improving global access to new medications.

5.4.2 *Improved Understanding of Customer Needs*

The pharmaceutical industry is at a precipice of significant change. The old product development models are being challenged, and new approaches are being adopted. In a time when total healthcare costs are being scrutinized by government and private payers alike, providers must maximize the quality of care by using

[9] Europe (Top 5) comprises France, Germany, Italy, Spain, and United Kingdom.

[10] Launch delay is a combination of submission delay by the sponsor company and the time it takes to review the product dossier by the regulatory agency.

interventions that provide the most value while minimizing cost. While drug prices seem to be continually rising, many health plans are shifting more of the burden of cost-sharing to patients. As a result, the patient is becoming more of a "consumer" in the traditional sense. This shift demands that the pharmaceutical industry begin to use the types of consumer research that are a mainstay of other industries in which the actual value of a product is defined by those who will ultimately use it. Notably, in these industries, the input of consumers is sought throughout development, from product conception through market entry and beyond. Developing a treatment that can genuinely help improve patients' lives should be rooted in a firm understanding of the challenges they face in their daily lives, their needs, and the trade-offs they are willing to make to gain relief (Lowe et al., 2016).

While this was true even before COVID-19, the pandemic and the ensuing economic downturn have brought about a sea change in the way organizations look at the road ahead. Pharmaceutical companies are up against a fresh set of challenges even as the pandemic has spurred the adoption of technologies. Based on a research article by Deloitte, pharmaceutical executives expect the pandemic to accelerate the research, manufacturing, and marketing of drugs to treat and prevent the disease. It will also accelerate investments in a few areas, such as digital technologies (Ford et al., 2020a). There are significant changes in consumer attitudes and behavior, as shown in Fig. 5.14, which compiles the results of Deloitte's survey.

Consumers are changing their attitudes in numerous ways—from increased use of technology and willingness to share data to their interest in using tools to make more informed decisions about prescription drugs and care. Outside the pharmaceutical industry, however, engagement with patients is increasing. Patients have been invited by regulators such as the FDA, the European Medicines Agency, and the National Institute for Health and Care Excellence to provide their perspectives and advice during decision-making. In the Prescription Drug User Fee Act (PDUFA[11]) V spanning fiscal years 2013 to 2017, the FDA launched the Patient-Focused Drug Development initiative designed to gain the patient perspective in specific disease areas and, in turn, to more effectively inform the FDA's evaluation of the risks and benefits of new therapies (Lowe et al., 2016). In fact, the major emphases of PDUFA VI[12], which track more or less with provisions in the twenty-first-century Cures bill, are enhancement of patient involvement in the drug approval process and the FDA's allowance of biomarkers, alternative clinical trials, and real-world evidence (Barlas, 2017).

---

[11] PDUFA was a law passed by the US Congress in 1992 which allowed the FDA to collect fees from drug manufacturers to fund the new drug approval process. The Act provided that the FDA was entitled to collect a substantial application fee from drug manufacturers at the time an NDA or BLA was submitted, with those funds designated for use only in CDER or CBER drug approval activities. In order to continue collecting such fees, the FDA is required to meet certain performance benchmarks, primarily related to the speed of certain activities within the review process.

[12] On August 18, 2017, President Donald Trump signed into law the Food and Drug Administration Reauthorization Act (FDARA), which includes the reauthorization of PDUFA through September 2022. PDUFA VI will provide for the continued timely review of new drug and biologic license applications.

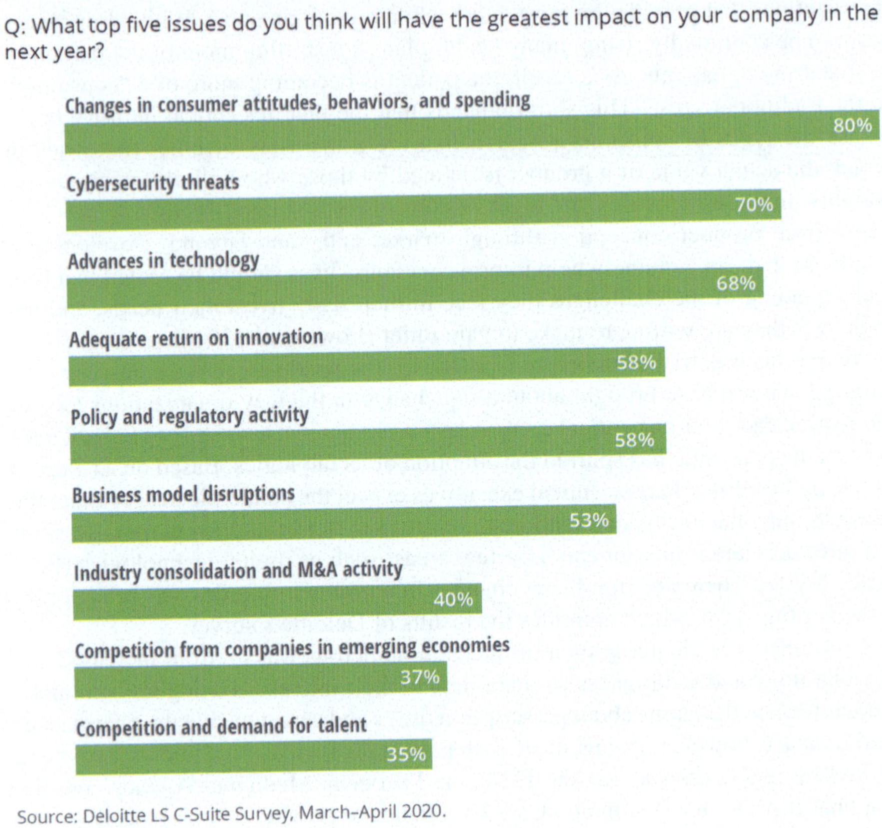

Fig. 5.14 Survey results of potential issues impacting pharmaceuticals. (Ford et al., 2020a)

One way pharmaceutical companies can continue to incorporate customer feedback into product development is by making sure that a drug is not only safe and effective but also that it truly meets the needs of patients. Without including patients in this conversation early on, it will be difficult to fully understand those needs and the trade-offs that patients are willing to make. Patients' perspectives on disease burden, treatment experience, risk-benefit assessments, and unmet needs may be incorporated early, informing the drug discovery phase or during the shaping of the target product profile (TPP). Among other things in a TPP, the patient voice can inform product attributes such as the route of administration, dosing, efficacy, acceptable levels of adverse events, and possibly even willingness to pay. Such an approach will lead to developing a patient-centric TPP that reflects patients' burden of disease, perspective on disease severity, and experience with existing treatment options. Companies should seek patient input in developing the TPP and continually engage patients to update this throughout the drug development process. This will help the organization identify suitable criteria and metrics in shaping their products, which, in turn, will steer drug development toward medications that address how patients function and feel, including their quality of life and unmet medical needs (Lowe et al., 2016).

Patient-reported outcome (PRO) studies aim to capture patient feedback on the perception and experience with drug treatment in the relevant disease, an emerging area of focus that will increasingly provide greater insight into product design (Stegemann et al., 2016). One such area is the development of medications for special patient populations (SPP) such as pregnant women, pediatrics, critically ill, or geriatrics. Historically, the pharmaceutical industry's role has been to develop science and medicines to prevent or treat disease for many patients. Patients in clinical trials during drug development are selected based on strict inclusion and exclusion criteria. As discussed in Chap. 3, unfortunately, this approach to clinical trial design leaves out SPPs.

Consequently, for the majority of drugs on the market, efficacy and safety are understudied and poorly characterized in SPP. As a result, dosing regimens are not selected for optimal efficacy or minimal toxicity in these patients. Also, SPP is often smaller than reference populations. Especially in children, practical and ethical constraints also limit the number of patients that can be included in studies and the number of invasive samples that can be obtained per patient. Moreover, studies in special populations are generally performed during routine clinical practice, which may negatively impact the information content of the obtained data. As a result, data obtained in special patient populations requires advanced methods with increased flexibility and statistical power to bring sufficiently high evidence from the pharmacokinetic and pharmacodynamics data to derive drug dosing regimens (Krekels et al., 2017). Most of the products for special patient populations require reformulation or adaptation of new drug delivery technologies.

5.4.3 *Reduction in Product Recalls and Manufacturing Issues*

As discussed before, a major reason for the federal role in regulating pharmaceutical quality is that primary customers (patients, caregivers, and health professionals) are often not able to assess the drugs' quality independently. The historical literature on drug regulation verifies this, telling a story of tragedies that occurred when unsuspecting health professionals treated patients with contaminated or improperly labeled drugs and of the subsequent laws enacted to prevent a recurrence. These laws generally increased federal government oversight of drug manufacturing operations (Fig. 5.15). From what happened in the case of a few pharmaceutical products and their subsequent enforcement efforts by the government, it is safe to say that our modern laws and quality check methodologies are designed to avoid repeating these tragic events. With the amount of oversight in the production of modern pharmaceutical products, it would be a fair assumption that quality-related problems do not persist anymore in pharmaceuticals. However, this is not entirely true!

The pharmaceutical industry has recently faced several challenges associated with using outdated development and manufacturing technologies. The return on

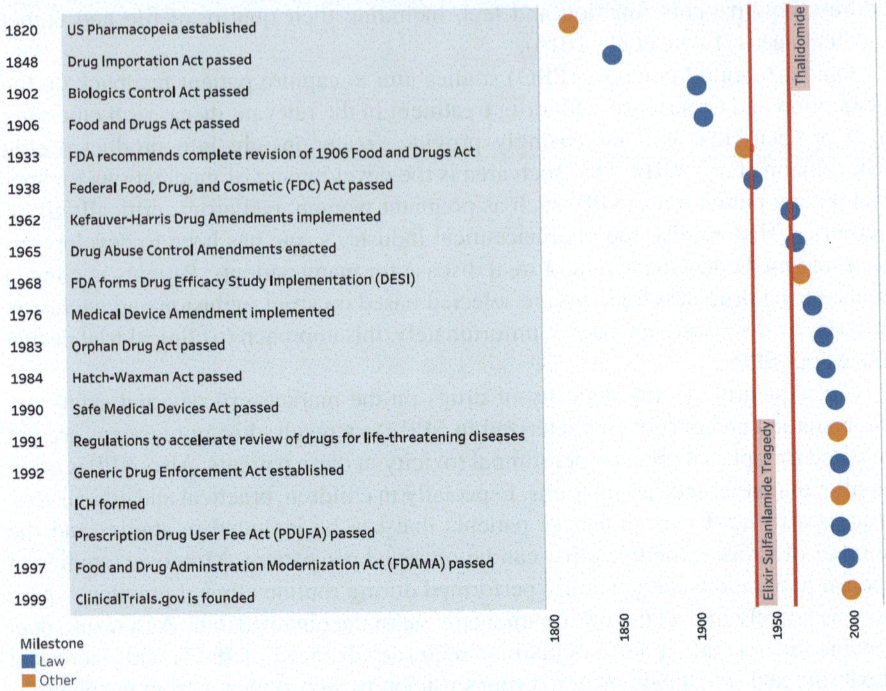

Fig. 5.15 Timeline of key laws and regulations for drugs and devices in the United States (1820–2000). (Reprinted with permission from Elsevier)

investment in research and development has been shrinking, and, at the same time, an alarming number of shortages and recalls for quality concerns have been registered. A key metric to evaluate the current pharmaceutical manufacturing and quality landscape is to monitor the number of drugs recalled in a given year. Drug recalls are the most effective way to protect the public from a defective or potentially harmful product. A recall is a voluntary action a company takes at any time to remove a defective drug product from the market. Drug recalls may be conducted on a company's initiative or by FDA request. FDA's role in a recall is to oversee a company's strategy and assess the recall's adequacy and classification. FDA maintains a database of the drugs it recalls and lists the reasons for the recall.

As per this database, an average of 44 marketed drugs per year were recalled from 2020 to 2023 (Fig. 5.16) for a myriad of reasons as shown in Fig. 5.17 for 2018. Most recall scenarios concerned chemistry, manufacturing, and controls (CMC). Therefore, more work must be done to improve the manufacturing quality of pharmaceutical products. In addition, a few key questions can be asked: (1) why are so many drugs being recalled annually? (2) what does this recall mean for the future of pharmaceutical quality? Moreover, (3) what can be done to reduce the number of product recalls?

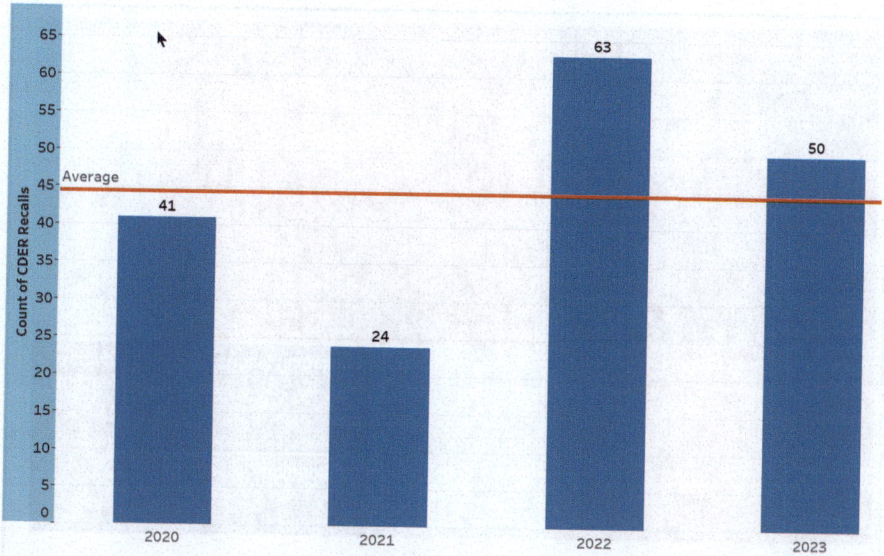

Fig. 5.16 Drug products recalled by FDA (2020–2023)

One way the pharmaceutical industry has been responding to these issues is through a technological modernization of development and manufacturing, under the support of initiatives and activities such as quality-by-design (QbD), process analytical technology (PAT), and pharmaceutical emerging technology (Destro & Barolo, 2022).

5.5 Opportunities to Extend Post-Peak Returns

Ensuring long-term profitability and revenue upon launching new drugs raises serious concerns for the involved parties in the pharmaceutical industry. In many industries, the loss of patent-protected exclusivity is known to be followed by severe losses in sales and profit to incumbent companies. When patent protection expires, generic manufacturers enter the market with drugs equivalent to the innovator's drug but typically at a significantly lower price. The impact of the loss of patent-protected exclusivity is shown in Fig. 5.18, where the sales dynamics of Pfizer's Lipitor®, formerly known as the best-selling prescription drug in history, show a dramatic decrease in sales once the period of market exclusivity is exhausted in 2012 (Song & Han, 2016).

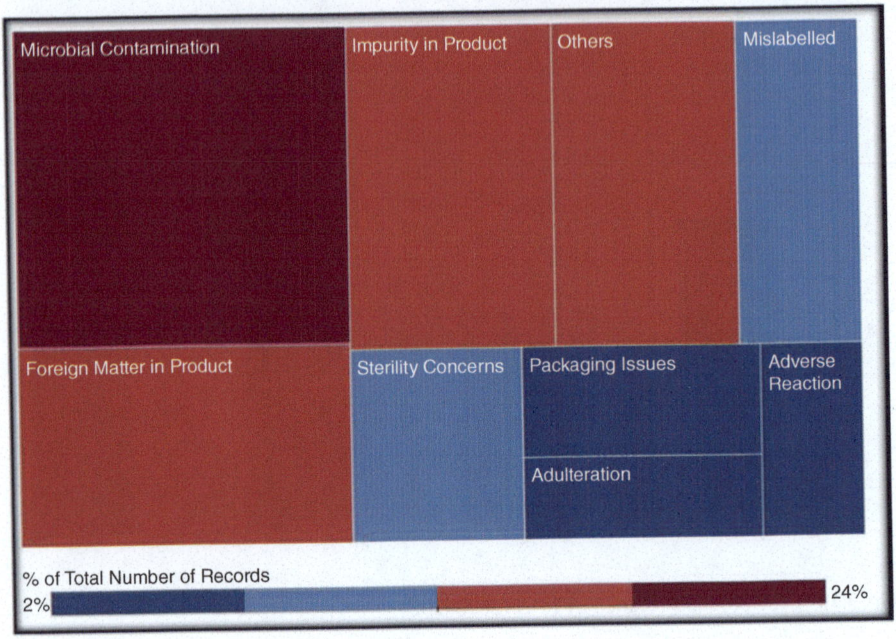

Fig. 5.17 Reason for drug products recalled by FDA in 2018. (Mittal, 2019) (Reprinted with permission from Elsevier)

Research-driven pharmaceutical companies can employ a range of strategies to extend their existing patent protection on therapeutically active substances or pharmaceutical composition, thus maximizing the commercial value of the product and retaining its market share. A detailed evaluation of these strategies is provided by (Song & Han, 2016), and a few technology-based strategies are discussed below. It is important to note that these technical strategies should complement any market protection actions the company undertakes to defend against competitive sales erosion. For example, in addition to the technical strategies, the company may also employ concurrent marketing strategies to maximize sales (such as a direct-to-consumer advertisement), licensing or selling of the exclusive rights to generic manufacturers, manufacturing branded generics, and competitive drug pricing strategies upon patent expiration (Zannou et al., 2009; Song & Han, 2016).

5.5.1 *Usage of Patent Law to Extend Protection*

Two types of patent protection exist for innovative pharmaceutical companies. The first are traditional patents granted and processed by the United States Patent and Trademark Office. The typical patent length is 20 years from the application's filing

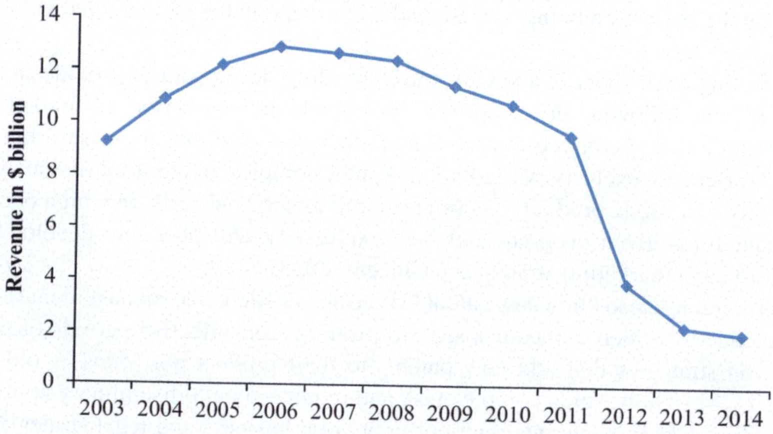

Fig. 5.18 Worldwide revenue profile of Pfizer's Lipitor®. (Song & Han, 2016)

date. Traditional patents can be filed anywhere along the development lifeline of a drug, and they can encompass a wide range of claims. The second type of protection is regulatory and granted by the FDA on approval of a New Chemical Entity (NCE) or biologic. This type of protection is termed "exclusivity" and could run concurrently with traditional patent protection. Exclusivity is designed to promote a balance between new drug innovation and generic drug competition.

Marketing exclusivity is granted only upon approval of a drug by the FDA and only when statutory requirements are met. Exclusivity puts into place a period of time during which no other applications can be accepted and approved for the same active ingredient. This means that other manufacturers that may wish to develop alternative formulations or generic versions of the drug will not be able to have their products approved during the exclusivity period. The length of the exclusivity period depends upon the type of exclusivity that has been granted. For an NCE, exclusivity is granted for a period of 5 years from the date of FDA approval. Orphan (for rare diseases and disorders) and pediatric drugs are granted longer exclusivity periods because of their limited use. For example, orphan drugs[13] are granted 7 years of exclusivity protection. Similarly, developing a product for the pediatric population provides the innovator's company with an additional 6 months of exclusivity. There are several types of marketing exclusivity, varying both in duration and the statutory requirements that must be met. Some are based on product classification,

[13] FDA's designation of an orphan drug is a drug that treats a disease or condition that affects less than 200,000 people in the United States, or affects more than 200,000 people in the United States but there is no reasonable expectation that the cost of developing and making the drug would be recovered.

others on the indication being treated, and still others on the intended patient population (Table 5.6).

Marketing exclusivity is a key incentive for drug developers. It provides a fixed period of time following drug approval, during which the sponsor can market their drug without direct competition from manufacturers of duplicate or reformulated products. Because exclusivity regulations can be complex and multiple exclusivities may apply to a single product, it is important to understand early on which ones are important for a given program and how exclusivity will play into development, regulatory, and marketing strategies (Allucent, 2023).

Evergreening, also known as patent "layering" or life-cycle management (LCM) is a practice by which companies seek to prolong their effective periods of patent protection strategies that add new patents to their product portfolios as old ones expire. LCM is, however, a complex task and requires a multidisciplinary approach. Successful LCM projects integrate not only good business and legal strategies but also good science and technologies. In fact, for timely implementation of the LCM strategy, activities are, in most cases, started even before the innovative product is filed with the regulatory agencies. There are many excellent cases where a product's life has been prolonged beyond the original patent expiration date, resulting in substantial financial returns or gains. One of the most commonly used LCM strategies includes brand protection strategies through salts, crystal forms, and formulations to extend time on the market without generic competition. When the first drug substance patent is filed, it may not cover all potential salts and polymorphs. Sometimes, a new salt or polymorph may have some advantage that can be patented separately. If used adequately, these new patents may provide additional life for the specific product. The same applies to new formulation patents. They may also include patents on reformulated products (Zannou et al., 2009).

5.5.2 *Continued Innovation Through Product Line Extensions*

The basic principle of 'innovation strategy' is to outpace the competition with a new move on the perceived value of a product or business model. In this sense, product-related innovation strategies comprise product-line extensions, approval of new indications, the introduction of follow-on products, and Rx-to-OTC switches, which all build on the company's knowledge assets and assets complementary to them. Product-line extensions refer to a variation of the existing products by proactive measures aimed at advancing them. Superior formulations of the known compound can promote patient compliance through reduced dosing, new route of administration, or improved compatibility (fewer side effects). Similarly, switching the prescription drug (Rx) to an over-the-counter (OTC) drug may provide a new venue for sustained revenue through expanding the market segment. The rationale underpinning the OTC switch is that patients with positive experiences with the branded drug are potentially more brand-loyal regarding product choice (Song & Han, 2016).

Table 5.6 Types of marketing exclusivity granted by FDA

Exclusivity type	Exclusivity period	Details
Biologic exclusivity	12 years	Biologic exclusivity conveys 12 years of total market protection. This includes 4 years before an application for a biosimilar may be submitted to the FDA and 8 additional years before an application may be approved
Orphan drug exclusivity (ODE)	7 years	A new brand-name drug for a disease or condition that affects fewer than 200,000 people in the United States (or that affects more people but for which the drug company still has no hope of covering the development costs) has a seven-year exclusivity
New chemical entity (NCE) exclusivity	5 years	In most cases, a brand-name drug with a new active moiety has a five-year exclusivity
Qualified infectious disease product (QDIP) exclusivity	5 years (added to any existing exclusivity)	Certain new antibiotic drugs for specific infectious diseases may be eligible for a five-year exclusivity, which is added on to any other exclusivities for that drug
Clinical investigation exclusivity (CIE)	3 years	A brand-name drug with an active ingredient that has been approved before may be awarded a three-year exclusivity in certain circumstances, such as if a new way of delivering the active ingredient is proposed (for example, a tablet rather than a liquid) or a different disease or condition the drug can treat is identified. To get this approval, the drug company must conduct new clinical studies in humans
Pediatric exclusivity (PED)	6 months (added to any existing exclusivity)	A brand-name drug for which the sponsor has done pediatric studies (in response to a written request from the FDA) may be eligible for a six-month exclusivity, which is added on to any other exclusivities or patents for that drug
Generic drug exclusivity (GDE)	180 days	The hatch-Waxman amendments provides an incentive of 180 days of market exclusivity to the "first" generic applicant who challenges a listed patent and runs the risk of having to defend a patent infringement suit

Adapted from Allucent (2023) and FDA (2023)

5.6 Summary—Enhancing the Value Proposition in Healthcare

Pharmaceutical development is a highly complex and innovative science. The current state of pharmaceutical development has seen its fair share of evolution in drug discovery, clinical trials, product design, and manufacturing. While this has helped us deliver more effective and better-tolerant products with fewer quality issues, the current state makes a limited effort to integrate economics into product design and development. This minimal integration of financial valuation elements into product development, unfortunately, hurts the sustainability of the health system of which these products will be a part in the future (Patel, 2017). A desirable future state integrates fundamentals of economics in product design and development so that the decision-making is parameterized, the cost of goods can be lowered, wastage can be reduced, patient-centricity is enhanced, manufacturing/distribution efficiencies can be gained, and life cycle management strategies are incorporated to maintain the commercial value of the innovator product. The financial benefits of such an approach could allow these savings to be passed on to the stakeholders and improve the value proposition of pharmaceuticals, which is critical to maintaining the innovation potential. ***How can such a desired state be accomplished for various pharmaceutical products?***

References

Allucent. (2023). *Types of marketing exclusivity in drug development*. Retrieved December 1, 2023 from https://www.allucent.com/resources/blog/types-marketing-exclusivity-drug-development#:~:text=Biologic%20exclusivity%20conveys%2012%20years,an%20application%20may%20be%20approved

Attene-Ramos, M. S., Austin, C. P., & Xia, M. (2014). High throughput screening. In P. Wexler (Ed.), *Encyclopedia of toxicology* (3rd ed., pp. 916–917). Academic. https://doi.org/10.1016/B978-0-12-386454-3.00209-8

Barlas, S. (2017). Drug companies breathe easier As PDUFA VI Heads for Approval. *Pharmacy and Therapeutics, 42*(6), 357.

CADLOG. (2023). *Cost optimization in electronics*. https://www.cadlog.com/cost-optimization-in-electronics/

CBO. (2021). *Research and development in the pharmaceutical industry*. Retrieved from https://www.cbo.gov/publication/57126

Chiva-Gomez, R. (2004). Repercussions of complex adaptive systems on product design management. *Technovation, 24*(9), 707–711. https://doi.org/10.1016/S0166-4972(02)00155-4

Chow, S. C., & Corey, R. (2011). Benefits, challenges and obstacles of adaptive clinical trial designs. *Orphanet Journal of Rare Diseases, 6*, 79. https://doi.org/10.1186/1750-1172-6-79

Desborough, M. J. R., & Keeling, D. M. (2017). The aspirin story—From willow to wonder drug. *British Journal of Haematology, 177*(5), 674–683. https://doi.org/10.1111/bjh.14520

Destro, F., & Barolo, M. (2022). A review on the modernization of pharmaceutical development and manufacturing—Trends, perspectives, and the role of mathematical modeling. *International Journal of Pharmaceutics, 620*, 121715. https://doi.org/10.1016/j.ijpharm.2022.121715

Doytchinova, I. (2022). Drug design—Past, present, future. *Molecules, 27*, 1496. https://doi.org/10.3390/molecules27051496

EFPIA. (2022). *The pharmaceutical industry in figures*. https://www.efpia.eu/media/637143/the-pharmaceutical-industry-in-figures-2022.pdf

FDA. (2001). *Guidance for industry: E 10 choice of control group and related issues in clinical trials*. Retrieved from https://www.fda.gov/downloads/guidances/ucm073139.pdf

FDA. (2004). *Pharmaceutical cGMPs for the 21st Century – A Risk-Based Approach, Final Report – Fall 2004*. Retrieved from https://www.fda.gov/downloads/drugs/developmentapprovalprocess/manufacturing/questionsandanswersoncurrentgoodmanufacturingpracticescgmpfordrugs/ucm176374.pdf

FDA. (2018a). *2017 new drug therapy approvals*. Retrieved from https://www.fda.gov/downloads/AboutFDA/CentersOffices/OfficeofMedicalProductsandTobacco/CDER/ReportsBudgets/UCM591976.pdf

FDA. (2018b). *Drug study designs—Information sheet*. Retrieved from https://www.fda.gov/RegulatoryInformation/Guidances/ucm126501.htm

FDA. (2019). *Adaptive designs for clinical trials of drugs and biologics guidance for industry*. Retrieved from https://www.fda.gov/media/78495/download

FDA. (2023). *Exclusivity and generic drugs: What does it mean?* Retrieved from https://www.fda.gov/files/drugs/published/Exclusivity-and-Generic-Drugs%2D%2DWhat-Does-It-Mean-.pdf

Fischer, M., Leeflang, P. S. H., & Verhoef, P. C. (2010). Drivers of peak sales for pharmaceutical brands. *Quantitative Marketing and Economics, 8*(4), 429–460. https://doi.org/10.1007/s11129-010-9089-5

Ford, J., Blair, A., Naaz, B., & Overman, J. (2020a). Deloitte insights: Biopharma leaders prioritize R&D, technological transformation, and global market presence. *Deloitte*. https://www2.deloitte.com/us/en/insights/industry/life-sciences/pharmaceutical-industry-trends.html

Ford, J., Fezza, T., Elsner, N., & Arora, A. (2020b). *Key factors to improve drug launches*.

Gautam, A., & Pan, X. (2016). The changing model of big pharma: Impact of key trends. *Drug Discovery Today, 21*(3), 379–384. https://doi.org/10.1016/j.drudis.2015.10.002

Gaynes, R. (2017). The discovery of penicillin—New insights after more than 75 years of clinical use. *Emerging Infectious Diseases, 23*(5), 849–853. https://doi.org/10.3201/eid2305.161556

Hansen, K. R. N. (2014). *New product introduction in the pharmaceutical industry*. DTU Management Engineering. https://backend.orbit.dtu.dk/ws/portalfiles/portal/101637275/New_Product_Introduction_in_the_Pharmaceutical_Industry.pdf

Heracleous, L., Terrier, D., & Gonzalez, S. (2018). The reinvention of NASA. *Harvard Business Review*.

Hutchings, M. I., Truman, A. W., & Wilkinson, B. (2019). Antibiotics: Past, present and future. *Current Opinion in Microbiology, 51*, 72–80. https://doi.org/10.1016/j.mib.2019.10.008

ICH. (2009). *Pharmaceutical development: Q8(R2)*. Retrieved from https://www.ich.org/fileadmin/Public_Web_Site/ICH_Products/Guidelines/Quality/Q8_R1/Step4/Q8_R2_Guideline.pdf

Iuga, A. O., & McGuire, M. J. (2014). Adherence and health care costs. *Risk Management and Healthcare Policy, 7*, 35–44. https://doi.org/10.2147/RMHP.S19801

Kaitin, K. I. (2010). The landscape for pharmaceutical innovation: Drivers of cost-effective clinical research. *Pharm Outsourcing, 2010*.

King, J. (2017). *Fixed-dose combination products—Navigating the combination rule*. Retrieved October 7 from https://camargopharma.com/2017/03/fixed-dose-combination-products-navigating-the-combination-rule/

Krekels, E. H. J., van Hasselt, J. G. C., van den Anker, J. N., Allegaert, K., Tibboel, D., & Knibbe, C. A. J. (2017). Evidence-based drug treatment for special patient populations through model-based approaches. *European Journal of Pharmaceutical Sciences, 109s*, S22–s26. https://doi.org/10.1016/j.ejps.2017.05.022

Kwon, K. C., & Lee, C. (2017). Analysis of fixed-dose combination products approved by the US food and drug administration, 2010–2015: Implications for designing a regulatory shortcut to new drug application. *Therapeutic Innovation & Regulatory Science, 51*(1), 111–117. https://doi.org/10.1177/2168479016663263

Lowe, M. M., Blaser, D. A., Cone, L., Arcona, S., Ko, J., Sasane, R., & Wicks, P. (2016). Increasing patient involvement in drug development. *Value in Health, 19*(6), 869–878. https://doi.org/10.1016/j.jval.2016.04.009

Lublóy, Á. (2014). Factors affecting the uptake of new medicines: A systematic literature review. *BMC Health Services Research, 14*, 469. https://doi.org/10.1186/1472-6963-14-469

Makurvet, F. D. (2021). Biologics vs. small molecules: Drug costs and patient access. *Medicine in Drug Discovery, 9*, 100075. https://doi.org/10.1016/j.medidd.2020.100075

Mandal, S., Moudgil, M. N., & Mandal, S. K. (2009). Rational drug design. *European Journal of Pharmacology, 625*(1), 90–100. https://doi.org/10.1016/j.ejphar.2009.06.065

Mittal, B. (2019). How to integrate quality by efficient design (QbED). In *Product development*. Elsevier Academic Press.

Morris, L. S., & Schulz, R. M. (1992). Patient compliance—An overview. *Journal of Clinical Pharmacy and Therapeutics, 17*(5), 283–295. https://doi.org/10.1111/j.1365-2710.1992.tb01306.x

NASA. (2022). *About NASA*. Retrieved December 29, 2022 from https://www.nasa.gov/about/index.html

Pallmann, P., Bedding, A. W., Choodari-Oskooei, B., Dimairo, M., Flight, L., Hampson, L. V., Holmes, J., Mander, A. P., Odondi, L., Sydes, M. R., Villar, S. S., Wason, J. M. S., Weir, C. J., Wheeler, G. M., Yap, C., & Jaki, T. (2018). Adaptive designs in clinical trials: Why use them, and how to run and report them. *BMC Medicine, 16*(1), 29–29. https://doi.org/10.1186/s12916-018-1017-7

Park, J. H., & Shin, K. (2018). R&D project valuation considering changes of economic environment: A case of a pharmaceutical R&D project. *Sustainability, 10*(4), 993. https://www.mdpi.com/2071-1050/10/4/993

Patel, N. A. (2017). Fee-for-value in the pharmaceutical industry: A policy framework applying data science to negotiate drug prices. *Journal of Law and the Biosciences, 4*(1), 205–215. https://doi.org/10.1093/jlb/lsx002

Piantadosi, S. (2017). *Clinical trials: A methodologic perspective* (3rd ed.). Wiley.

Samanen, J. M. (2013). *The structure and business of bio/pharmaceutical companies including the management of risks and resources*. Elsevier.

Sarkis, M., Bernardi, A., Shah, N., & Papathanasiou, M. M. (2021). Emerging challenges and opportunities in pharmaceutical manufacturing and distribution. *Processes, 9*(3), 457. https://www.mdpi.com/2227-9717/9/3/457

Song, C. H., & Han, J. W. (2016). Patent cliff and strategic switch: Exploring strategic design possibilities in the pharmaceutical industry. *Springerplus, 5*(1), 692. https://doi.org/10.1186/s40064-016-2323-1

Stegemann, S., Ternik, L., Onder, G., Khan, M., & van Riet-Nales, D. (2016). Defining patient centric pharmaceutical drug product design. *The AAPS Journal, 18*, 1047–1055. https://doi.org/10.1208/s12248-016-9938-6

Swann, J. P. (1995). The evolution of the American pharmaceutical industry. *Pharmacy in History, 37*(2), 76–86.

Thind, M., & Kowey, P. R. (2020). The role of the food and drug administration in drug development: On the subject of Proarrhythmia risk. *Innovations in Cardiac Rhythm Management, 11*(1), 3958–3967. https://doi.org/10.19102/icrm.2020.110103

Van Alsten, S. C., Harris, J. K. (2020). Cost-related nonadherence and mortality in patients with chronic disease: A multiyear investigation, national health interview survey, 2000–2014. *Preventing Chronic Disease* 17, 200244. http://dx.doi.org/10.5888/pcd17.200244

Viswanathan, M., Golin, C. E., Jones, C. D., et al. (2012). Interventions to improve adherence to self-administered medications for chronic diseases in the United States: A systematic review. *Annals of Internal Medicine, 157*(11), 785–795. https://doi.org/10.7326/0003-4819-157-11-201212040-00538

Vondeling, G. T., Cao, Q., Postma, M. J., & Rozenbaum, M. H. (2018). The impact of patent expiry on drug prices: A systematic literature review. *Applied Health Economics and Health Policy, 16*(5), 653–660. https://doi.org/10.1007/s40258-018-0406-6

White, B. (2013). *4 types of dose finding studies used in phase II clinical trials*. http://resources.rhoworld.com/blog/bid/273120/4-types-of-dose-finding-studies-used-in-phase-ii-clinical-trials

Wouters, O. J., McKee, M., & Luyten, J. (2020). Estimated research and development investment needed to bring a new medicine to market, 2009–2018. *JAMA, 323*(9), 844–853. https://doi.org/10.1001/jama.2020.1166

Yu, L. X., Amidon, G., Khan, M. A., Hoag, S. W., Polli, J., Raju, G. K., & Woodcock, J. (2014). Understanding pharmaceutical quality by design. *The AAPS Journal, 16*(4), 771–783. https://doi.org/10.1208/s12248-014-9598-3

Zannou, E. A., Li, P., & Tong, W.-Q. (2009). Chapter 40—Product lifecycle management (LCM). In Y. Qiu, Y. Chen, G. G. Z. Zhang, L. Liu, & W. R. Porter (Eds.), *Developing solid oral dosage forms* (pp. 911–921). Academic. https://doi.org/10.1016/B978-0-444-53242-8.00040-0

Zele, C. (2023). *Market access 101: Understanding the basics*. Retrieved October 28 from https://www.mmitnetwork.com/thought-leadership/market-access-101-understanding-the-basics/

Zhu, X., & Liu, B. (2020). Launch delay of new drugs in China and effect on patients' health. *Clinical Therapeutics, 42*(9), 1750–1761.e1757. https://doi.org/10.1016/j.clinthera.2020.06.023

Chapter 6
Enhancing the Value Proposition of Small Molecule Pharmaceuticals

Any sufficiently advanced technology is indistinguishable from magic

Arthur C. Clarke

Contents

Abstract Small molecules have provided medical breakthroughs for human diseases for over a century. As in any R&D endeavor, science gains momentum with time, and the accretive knowledge of research is due to the collective efforts of many talented scientists and engineers. However, as new knowledge is gathered, it is equally important to integrate it with existing paradigms to make our processes and products more cost-efficient and robust. Small molecule-based drug development is a prime example of an area that can benefit tremendously from such an inclusive and integrative approach. During the design of a dosage form, formulators need to take into consideration challenges associated with stability, bioavailability, manufacturability, acceptability, and business acuity. A formulator must aim to address all of these critical areas, no matter the drug product's development stage.

B. Mittal, *Sustainable Global Health Systems and Pharmaceutical Development*, AAPS Advances in the Pharmaceutical Sciences Series 60,
https://doi.org/10.1007/978-3-031-50415-0_6

The failure to address any of these elements can lead to significant manufacturing and performance issues that may not be discovered till late in the development cycle. A formulator needs to realize that decisions made during the design phase of a product's development determine most of the costs of manufacturing that product. As the design and manufacturing processes become more complex and scale, the formulator will increasingly be called upon to make decisions that involve a significant investment of time, people, and money. Each of these decisions cannot be made in isolation, and a balance must be struck each time to ensure that none of the other design rules of stability, bioavailability, manufacturability, and acceptability are violated. Poor product design without the foresight of scale-up, marketing, and global compliance may lead to wasting precious resources that could have been invested in the company's pipeline. Embracing new manufacturing philosophies and technologies such as continuous manufacturing, additive manufacturing, artificial intelligence, and Six Sigma pharmaceutical quality can also enhance the value proposition of small molecule-based drug development.

Keywords Solid oral dosage drug development · Business acuity · Biopharmaceutical classification system · Developability classification system · Modeling and simulation · Continuous manufacturing · Additive manufacturing · Artificial intelligence

6.1 Introduction

It is a hallmark of our modern technological society that change is everywhere one looks. In just the last 100–150 years, we have had cars, airplanes, spaceships, computers, and numerous other devices, technologies, and scientific advances that not only have the potential to make our lives better but also could lead to irreversible damage to our planet. In an excellent article by Professor Max Roser,[1] the founder of Our World in Data, he postulates that it is easy to underestimate how much the world can change within a lifetime because of technology. Technology can change the world in ways that are unimaginable until they happen. A few generations ago, it would have been unthinkable how computers might change our lives. Similarly, there are technologies that are being developed today whose impact we cannot yet fathom, but they will be commonplace for a future generation. In an excellent visualization (Fig. 6.1) developed by Dr. Roser, we can appreciate the long-term perspective on the history of technology.

[1] Dr. Max Roser is the Professor of Practice in Global Data Analytics at the University of Oxford's Blavatnik School of Government, the Program Director of the Oxford Martin Programme on Global Development, and Co-executive Director of Global Change Data Lab, the non-profit organization that publishes Our World in Data.

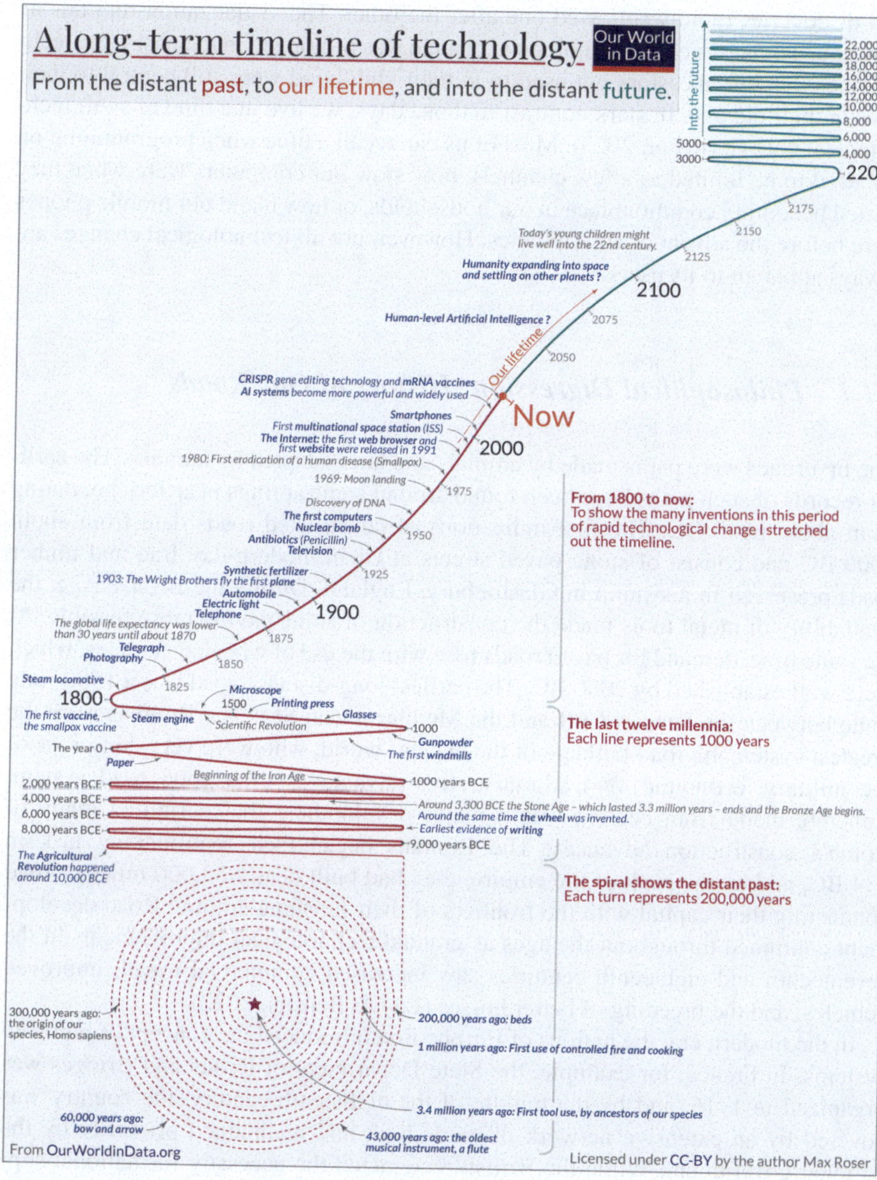

Fig. 6.1 A long-term timeline of technology. (Roser, 2023)

The timeline begins at the center of the spiral. The first use of stone tools, 3.4 million years ago, marks the beginning of this history of technology. Each turn of the spiral represents 200,000 years of history. It took 2.4 million years—12 turns of the spiral—for our ancestors to control fire and use it for cooking. However, since 1800, the pace of technological progress dramatically increased as many major inventions and discoveries rapidly followed one after the other. The visualization reminds us how unusual our time is. Technological change was prolonged in the past—the technologies that our ancestors got used to in their childhood were still central to their lives in their old age. In stark contrast to those days, we live in a time of swift technological change (Roser, 2023). Most of us can recall a time when programming on TV used to be limited to a few channels, how slow our computers were when they started becoming commonplace in our households, or how bland our mobile phones were before the advent of smartphones. However, not all technological changes are always apparent to its users.

6.1.1 Philosophical Digression: History of Our Roads

The first roads were paths made by animals and later adapted by humans. The earliest records of such paths have been found around some springs near Jericho, dating from about 6000 BC. The first indications of constructed roads date from about 4000 BC and consist of stone-paved streets at Ur in modern-day Iraq and timber roads preserved in a swamp in Glastonbury, England. During the Bronze Age, the availability of metal tools made the construction of stone paving more feasible. At the same time, demand for paved roads rose with the use of wheeled vehicles, which were well established by 2000 BC. The earliest long-distance road was a 1500-mile route between the Persian Gulf and the Mediterranean Sea. The Romans were the greatest systematic road builders of the ancient world, who were very conscious of the military, economic, and administrative advantages of a good road system. Concrete made from cement was a major development that permitted many of Rome's construction advances. The Romans began their road-making task in 334 BC, and by the peak of the empire, they had built nearly 53,000 miles of road connecting their capital with the frontiers of their far-flung empire. Road development continued throughout the ages as gradual technological improvements in the seventeenth and eighteenth centuries saw increased commercial travel, improved vehicles, and the breeding of better horses (Lay & Benson, 2022).

In the modern era, the nations of Europe first introduced the concept of highway systems. In France, for example, the State Department of Roads and Bridges was organized in 1716, and by the middle of the eighteenth century, the country was covered by an extensive network of roads built and maintained primarily by the national government. While the British recognized the necessity for national support of highways and a national system as early as 1878, it was the Ministry of Transport Act of 1919 that first classified the roadway system into 23,230 miles of Class I roads and 14,737 miles of Class II roads. In the United States, the mammoth U.S. Interstate Highway System (formally the National System of Interstate and Defense Highways) was developed in response to strong public pressures in the 1950s for a better road system. These pressures culminated in the establishment of the Clay Committee by President Dwight Eisenhower[2] in 1954. Following this committee's recommendations, the Federal Aid Highway Act and the Highway Revenue Act of 1956 provided funding for an accelerated program of construction. A federal gasoline tax was established, the funds from which, with other highway-user payments, were placed in a Highway Trust Fund. The federal-state ratio for funding construction of the Interstate System was changed to 90% federal and 10% state. It was expected that the system would be completed no later than 1971, but cost increases and planning delays extended this time by some 25 years. The system grew to a total length of more than 45,000 miles, connecting nearly all the major cities in the United States and carrying more than 20% of the nation's traffic on slightly more than 1% of the total road and street system (Lay & Benson, 2022). The US Interstate System has been called the "Greatest Public Works Project in History" that has transformed the nation. However, that is not the end of the story.

Civil engineers and urban planners realize that due to the increasing number of vehicles on roads at an alarming rate, the existing infrastructure and conventional methods of traffic management are becoming inefficient. There is a need to adapt the old to the changing new world order, where daily processes are driven with the help of innovative technologies. In that regard, researchers are evaluating cutting-edge technologies such as autonomous vehicles, connected vehicles, the Internet of Things (IoT), and big data analytics to usher in the future state of our highway system. For example, it is anticipated that autonomous vehicles will provide increased convenience and safety for travelers, improved traffic flow, highway capacity, fuel economy, better mobility for elderly and disabled people, and reduced costs, congestion, emissions, and road accidents. Similarly, big data analytics can help in better traffic monitoring systems, avoiding congestion, accident prevention, and real-time monitoring. Although there are challenges associated with how to make this type of new technology implementation cost-effective and secure in an existing infrastructure, a futuristic vision propels the efforts, and it is highly likely that these technological advancements can provide highway operators with a solution that might resolve unforeseeable challenges (Azmat et al., 2019).

6.1.2 Opportunity for Pharmaceutical Scientists: Keep Innovating and Integrating New Knowledge

Why is innovation and integration of new knowledge important? It all depends on the perspective of the reader. As in any area of research and development, science gains momentum with time, and the accretive knowledge of research is due to the collective efforts of many talented scientists and engineers. However, as new

---

[2] President Eisenhower gained an appreciation of the Reichsautobahn system, the first "national" implementation of modern Germany's Autobahn network, as a necessary component of a national defense system while he was serving as Supreme Commander of Allied Forces in Europe during World War II.

knowledge is gathered, it is equally important to integrate it with existing paradigms to make our processes and products more cost-efficient and robust. Small molecule-based drug development is a prime example of an area that can benefit tremendously from such an inclusive and integrative approach.

Throughout our collective history, efforts have been made to extend the quality of life of people suffering from various ailments. From time immemorial, plants have been used for medicinal purposes. Several ancient cultures, such as Egyptians, Sumerians, Chinese, Indians, and Greeks, wrote about plants and their medicinal uses. However, the discovery of alkaloids began with Friedrich Sertürner's isolation and discovery of morphine[3] as the hypnotic principle of opium in 1805, which was among the most significant therapeutic advances of the early nineteenth century. Sertürner's discovery stimulated a search for active principles in other medicinal plants, and eventually, this would be an enormous contribution to the development of the modern pharmaceutical industry (Swann, 1995).

In an interesting article written by Dr. John P. Swann of the History Office of FDA, he clarifies that imports from Europe were the principal source for drugs in pre-nineteenth century America. In Europe, at this time in history, even though mass production of certain drugs had been evident for at least three centuries, the beginning of a true pharmaceutical industry was still more than a century away (Swann, 1995). Meanwhile, in America, in 1818, pharmacists John Farr and Abraham Kunzi opened an establishment that soon produced quinine and other alkaloids in bulk. However, a lack of uniformity in medical practices, as well as a desire to demonstrate a young nation's independence from Europe, led to the development of the first US Pharmacopeia (USP).

The first USP was published on December 21, 1820 (USP, 2018). A few decades later, companies such as SmithKline and French, William S. Merrell, Parke-Davis, and Wyeth came into existence. In 1843, English inventor William Brockedon was granted a patent for a device capable of "shaping pills, lozenges and black lead by pressure in dies." This was the progenitor of the modern-day tablet press (LeDoux, 2016). By the Civil War (1861–1865), there was a growing presence of manufactured drugs. In Philadelphia alone, by 1882, there were 144 companies engaged in some pharmaceutical manufacture (Swann, 1995). Pharmaceutical drug regulation was established in 1906 (please see Table 4.1), followed by the rapid growth of pharmaceuticals with the institutionalization of science and research in drug manufacturing. A time chart of selected small-molecule drugs from Aspirin onward, demonstrating the broad scope of small molecules as therapeutics, is shown in Fig. 6.2. What the global pharmaceutical industry has achieved is truly remarkable!

[3] Friedrich Wilhelm Adam Sertürner (1783–1841), a German pharmacist, was the first to isolate morphine from opium. He called the isolated alkaloid "morphium" after the Greek god of dreams, Morpheus. He published a comprehensive paper on its isolation, crystallization, crystal structure, and pharmacological properties. It was not only the first alkaloid to be extracted from opium, but the first ever alkaloid to be isolated from any plant.

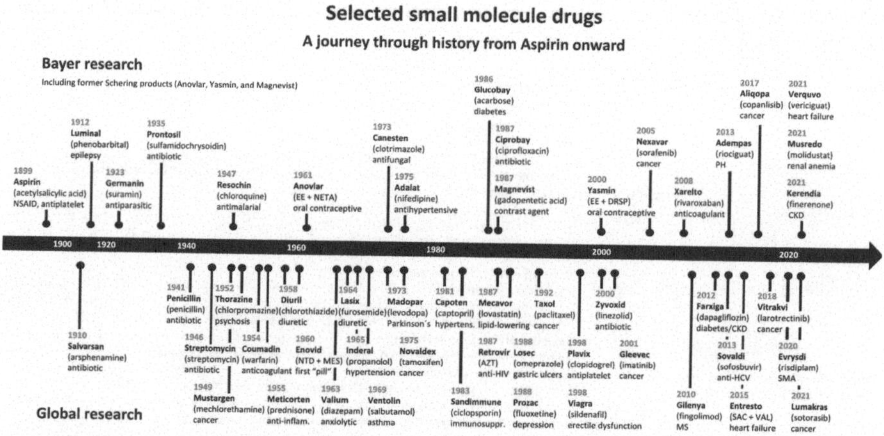

Fig. 6.2 Timeline of selected small-molecule drugs from aspirin onwards. (Beck et al., 2022) (Reprinted with permission from Elsevier)

For a casual observer, the humble tablet (or capsule) of today might look eerily similar to what it looked like even 150 years ago. So, what is different?

6.2 Small Molecules-Based Product Development

Small molecules have provided medical breakthroughs for human diseases for over a century. To a large part, their success as drugs has been due to their inherent properties, including their ability to cross biological barriers and to modulate an array of different biological targets. Oral bioavailability is a key feature of most small-molecule drugs, enabling standard oral delivery as a tablet. This convenience aspect is a huge advantage compared with biologics, even if the pharmacological effect size might be similar. Examples are the small-molecule inhibitors of hypoxia-inducible factor prolyl hydroxylase (HIF-PH), which can replace the recombinant biologic erythropoietin (EPO) in the treatment of renal anemia, thus replacing injection/infusion administration with a pill. Another important feature of small molecules is their modular structure and their ease of accessibility via chemical synthesis. This allows for rapid variation of the chemical structure and systematic improvement of their properties. In most cases, drug products contain the unmodified small molecule as the active ingredient. However, suitable reversible chemical modifications (prodrugs) can be designed to address potential shortcomings, such as low aqueous solubility. Finally, small molecules usually show high shelf-life stability and are compatible with most drug formulations and administration routes (Beck et al., 2022). Several more key characteristics of small molecules are summarized in Table 6.1.

In Sect. 5.3.1, an overview of rational drug design was presented. Rational drug design is the most brilliant and cheapest approach to drug discovery, and it begins

Table 6.1 Key properties of small molecules impacting their use in drug discovery (Beck et al., 2022) (Reprinted with permission from Elsevier)

Property	Small molecule
Biological target space	Very diverse: e.g., proteins (enzymes, ion channels, receptors, etc.), RNA, DNA, (glyco-)lipids Intracellular and extracellular targets
Functional interaction with target	Diverse: inhibitor, activator, antagonist, agonist, (allosteric) binder, etc. Reversible physical binding (i.e., no chemical modification of the biological target) versus irreversible chemical binding (i.e., permanent chemical modification of the biological target) Tunable properties
Target selectivity	Can be optimized to reach high target selectivity Tunable to address multiple (related) targets
Drug combinations	Can be combined with other drugs
Immunogenicity	Typically not
Permeability	Ability to cross biological barriers Good tissue penetration Cell permeable, intracellular targets accessible Tunable properties, e.g., to avoid crossing of blood–brain barrier
Metabolic stability	Tunable stability, from low to high Sufficiently high to allow oral application and prolonged systemic exposure
Application routes	No limitations; all application routes
Formulation and drug delivery	Compatible with standard and specialized formulations and drug delivery technologies From conventional instant-release tablets to sophisticated delivery systems (liposomes/exosomes, organic nanocrystals, nanoparticles, nanotubes, albumin conjugates, etc.)
Shelf-life stability	High stability in most cases Typically no demanding storage or transport conditions required
Chemical structure	Small to mid-size synthetic organic molecules Natural product or natural product-derived organic molecules Peptide or chemically modified peptides Macrocyclic and other shapes Variable complexity, both in terms of chemical structure and function (e.g., bifunctional drug conjugates) Modular structure, easily accessible for optimization of properties
Production by chemical synthesis	Mature technology, from laboratory to industrial scale Provides great flexibility, in terms of molecular structure and synthesis approaches Facilitates rapid analoging and derivatizations during drug discovery Cost of goods usually not critical
Driver of innovation	Expanding biological target space: e.g., protein surfaces in PPIs, RNA Orally bioavailable and cell-permeable peptides Protein degraders to induce removal of biological target, rather than its functional modulation Huge DNA-encoded small-molecule libraries for lead finding in drug discovery, making use of PCR technology for hit deconvolution Suitable for AI/ML-supported lead optimization in drug discovery

with the identification of a biological target (a biomacromolecule involved in the disease). A drug target is a biomolecule involved in signaling or metabolic pathways specific to a disease process (Mandal et al., 2009). Then, a ligand interacting with this macromolecule, known as a hit molecule, has to be discovered. It follows an iterative method of structure optimization until a compound is derived with optimal affinity, selectivity, non-toxicity, solubility, permeability, and bioavailability properties that are needed for a molecule to become a drug (Doytchinova, 2022). Once a target is identified, several aspects of the development of small molecules need to be examined. These aspects include but are not limited to the evaluation of binding scores (affinity/specificity), the balance between hydrophilicity/lipophilicity, ADME, electrophilic, nucleophilic, and radical attack (biodegradation), the toxicity of the parent small molecules, and products due to biotransformation in the different phases of metabolism, quantitative structure–activity relationship (QSAR), and quantitative structure–property relationship (QSPR) respectively (Mandal et al., 2009). This book does not discuss other aspects of drug development, such as candidate selection, drug substance development, non-clinical studies, and clinical studies. A critical aspect of the drug development process is drug product development, which is the most visible component of the medication for the patient. However, drug product development is just a small portion of the entire picture of the drug development process, and one must have an appreciation of all aspects of the drug development process.

6.2.1 Biopharmaceutical Classification System (BCS) and Developability Classification System (DCS)

All routes of administration require the absorption of drugs into the blood. Once the drug reaches the bloodstream, it partitions between the plasma and the red blood cells, the erythrocytes. Drugs in the plasma further partition between the plasma proteins (mainly albumin) and the plasma water. It is this free or unbound drug in plasma water, and not the drug bound to proteins, which can pass out of the plasma through the capillary endothelium and reach other body fluids and tissues, and hence the site(s) of action. A dynamic equilibrium exists between the concentration of the drug in the blood plasma and the drug at its site(s) of action. This is termed distribution, the degree of which will depend largely on the physicochemical properties of the drug, in particular its lipophilicity. In addition to absorption and distribution, other processes such as metabolism and elimination are also important in the understanding of biopharmaceutics. The study and characterization of the time course of drug ADME is termed pharmacokinetics (Fig. 6.3) and is used in the clinical setting to enhance the safe and effective therapeutic management of individual patients (Ashford & Taylor, 2022).

As shown in Fig. 6.3, the rate and extent of appearance of the intact drug in the systemic circulation depends on a succession of kinetic processes. The slowest step in this series, which is known as the rate-limiting step, controls the overall rate and extent of appearance of intact drug in the systemic circulation. The particular

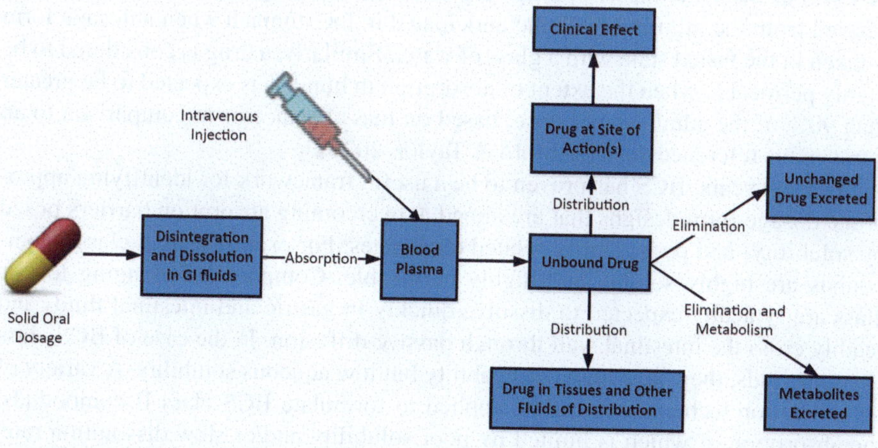

Fig. 6.3 Drug Absorption, Distribution, Metabolism, and Elimination (ADME). (Mittal, 2016) (Reprinted with permission from Elsevier)

Table 6.2 Biopharmaceutical classification system

Class	Solubility	Permeability
I	High	High
II	Low	High
III	High	Low
IV	Low	Low

rate-limiting step will vary from drug to drug. For example, for a drug that has very poor aqueous solubility, the rate at which it dissolves in the GI fluids is often the slowest step, and the bioavailability of that drug is said to be dissolution-rate limited. In contrast, for a drug that has high aqueous solubility, its dissolution will be rapid, and the rate at which the drug crosses the GI membrane may be the rate-limiting step. Other potential rate-limiting steps include the drug's rate of release from the dosage form, the rate at which the stomach empties the drug into the small intestine (gastric emptying), the rate at which the drug is metabolized by enzymes in intestinal mucosal cells during its passage through them into the mesenteric blood vessels, and the rate of drug's metabolism during its initial passages through the liver, often termed the first pass effect (Ashford & Taylor, 2022).

In 1995, pharmaceutical scientist proposed a scheme to classify drugs into four classes according to their solubility across the gastrointestinal pH range and their permeability across the gastrointestinal mucosa (Table 6.2). This has been known since as the biopharmaceutical classification system (BCS) (Amidon et al., 1995). BCS was originally proposed for the identification of immediate-release solid oral products for which in-vivo bioequivalence tests may not be necessary, but it is also useful to classify drugs and predict bioavailability issues that may arise during the various stages of the development process (Amidon et al., 1995).

According to this scheme, a drug is considered to be highly soluble where the highest dose strength is soluble in 250 mL or less of aqueous media over a pH range of 1–7.5 at 37 °C. Otherwise, drugs are considered poorly soluble. The volume is derived from the minimum volume anticipated in the stomach when a dosage form is taken in the fasted state with a glass of water. Similarly, a drug is considered to be highly permeable when the extent of absorption in humans is expected to be greater than 90% of the administered dose, based on mass balance or in comparison to an intravenous reference dose (Ashford & Taylor, 2022).

Over the years, BCS has proven to be a useful framework for identifying appropriate dosage form designs that are aimed at overcoming absorption barriers posed by solubility- and permeability-related challenges. For example, BCS class I compounds are highly soluble and highly permeable. Compounds belonging to this class are typically expected to dissolve quickly in gastric and intestinal fluids and readily cross the intestinal wall through passive diffusion. In the case of BCS class II compounds, they have high permeability but low aqueous solubility. A variety of solubilization technologies can be applied to formulate BCS class II compounds, the absorption of which is limited by poor solubility and/or slow dissolution rate. Overall, BCS class plays a significant role in choosing the path that will be taken when designing a dosage form (as shown in Fig. 6.4). Clearly, not all pathways lead

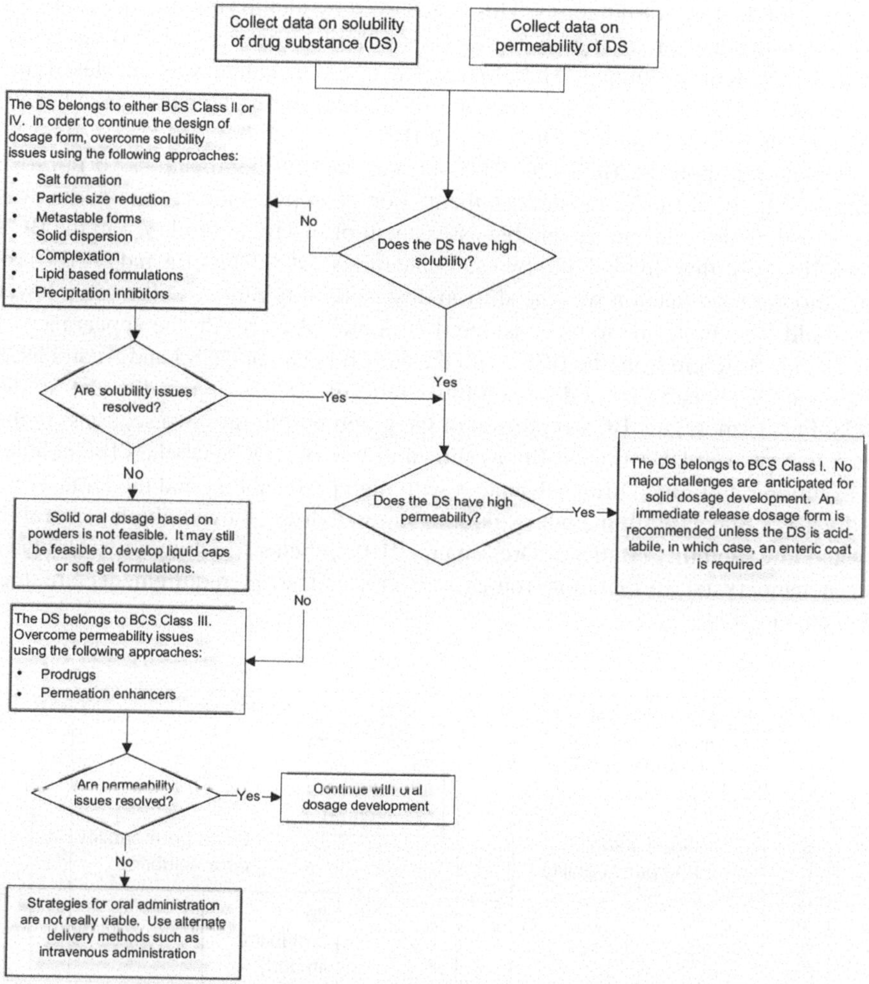

Fig. 6.4 Integration of BCS with decision-making process of developing dosage form design. (Mittal, 2016) (Reprinted with permission from Elsevier)

to the oral administration route, and even then, some of the oral dosage routes may not be feasible for tableting or dry powder encapsulation.

Since the introduction of the BCS, the solubility landscape of marketed drugs continues to evolve. It is conservatively estimated that approximately 40–60% of approved drugs suffer from poor solubility (BCS Class II or BCS Class IV). The same assessment applied to drug candidates in development reveals that 60–90% of drugs are classified as poorly soluble (Göke et al., 2018). Therefore, there is a high risk of failure due to poor performance and poor absorption in the gastrointestinal tract. In recent years, a developability classification system (DCS) has been proposed that modifies the BCS system to assist formulators in the development of poorly soluble drug formulations. This is achieved by identifying the root cause for low solubility and providing strategies for molecules that are either dissolution-limited or solubility-limited. Therefore, DCS is a convenient way of classifying compounds in a way that is more meaningful in determining human in vivo behavior and sensitivities (Butler & Dressman, 2010).

Expanding upon the BCS, the DCS introduced key modifications to improve applicability to formulation development. For example, biorelevant media was introduced to provide a more reliable assessment of in-vivo solubility, and the BCS Class II was further divided into two sub-categories: DCS Class IIa and DCS Class IIb. Another modification was the shift in dose solubility ratio, resulting in a lower threshold for a molecule to be considered "soluble". As a result, the appearance of the DCS is different from the BCS, with the cut-off between DCS I and III and DCS IIa/b and IV appearing at a dose-solubility ratio of 500, as opposed to 250 in the BCS. Furthermore, the BCS II portion of the graph is split into two sections[4] in the DCS to represent DCS IIa and IIb, as shown in Fig. 6.5. DCS subclass IIa includes dissolution rate-limited drug substances with high permeability and moderate solubility. DCS Subclass IIb includes solubility-limited drugs showing high permeability and low solubility (Butler & Dressman, 2010). Subclass IIa drugs are candidates for a nanocrystal formulation route if several industrial requirements are met (Uhlemann et al., 2021).

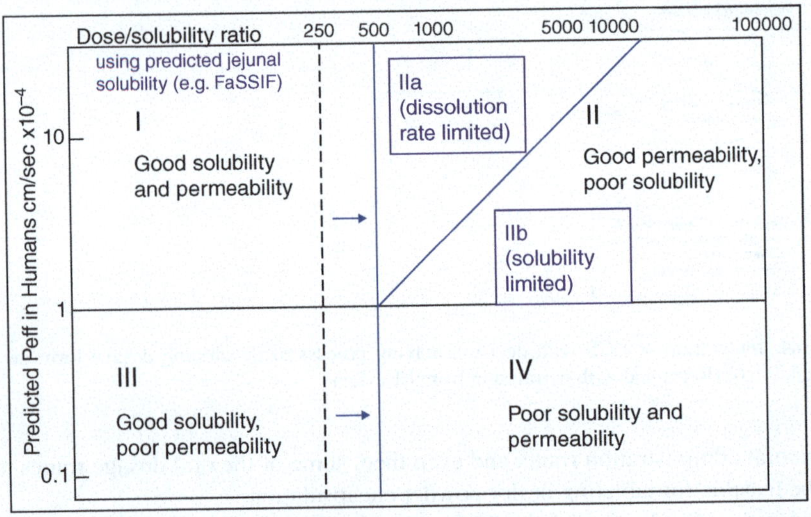

Fig. 6.5 Modification of BCS into DCS (shown in blue). (Butler & Dressman, 2010) (Reprinted with permission from Elsevier)

[4] The line which separates the two sub-categories is the solubility-limited absorbable dose (SLAD) line. The SLAD line uses a ratio of solubility and permeability to determine if an increase in dissolution rate will have a measurable impact on overall absorption. Based on this ratio, a drug molecule is placed either above or below the SLAD line. Molecules above the line are DCS Class IIa and the ones below the SLAD line are DCS Class IIb.

6.2.2 *Typical Solid Oral Dosage Product Development*

As people working in the field of Formulation Sciences would testify, Formulation Sciences is an amalgamation of numerous concepts developed in physical pharmacy, chemistry, material sciences, biopharmaceutics, and engineering. It is only on relatively few occasions that a drug, as such, may be directly administered to a patient. Usually, the medicament must be formulated with various excipients to ensure its intended performance. There are a few tried and trusted approaches that are typically followed during the design of a dosage form. These considerations are around stability, bioavailability, manufacturability, acceptability, and business acuity. These are discussed in detail in Table 6.3. It is imperative that a formulator should aim to address all of these critical areas no matter what stage of development the drug product may be in. The failure to address any of these elements can lead to major manufacturing and performance issues that may not be discovered till late in the development cycle.

Non-clinical safety information is important for estimating an initial safe starting dose for human trials and identifying parameters for clinical monitoring for potential adverse effects. It is very important to estimate a safe starting dose at a level that is not too low such that dose optimization is prolonged, as well as not too high to cause serious toxicity. Extrapolation of doses from animals to humans is based on multiple assumptions about the compound's behavior across species. Different methods may yield widely varying results, and an approach that has worked well for one compound may not be appropriate for another. There are multiple preclinical doses that are used as a basis for estimating the starting dose. One of the most commonly used bases is the no observed adverse effect level (NOAEL), defined as the highest dose at which no statistically significant and/or biologically relevant adverse effect is observed. Typically, one of the most commonly used methods for selecting the starting human dose involves the conversion of NOAEL to the human equivalent dose (HED) using appropriate scaling factors, followed by applying a safety factor (Mahmood et al., 2003).

From a formulator's perspective, the estimation of the initial safe starting point is an important design parameter as it determines the target product profile of the drug product and sets the strategy that determines the drug load, selection of excipients, and manufacturing process. It is very important for a formulator to understand the drug substance properties in order to develop a robust strategy. As a starting point, evaluating the BCS classification of the drug substance gives an understanding of which particular rate-limiting property of the drug substance needs to be targeted: its solubility, permeability, or both. An understanding of the BCS classification also inadvertently dictates the choice of excipients that will be made in order to target the rate-limiting property. It is also important to understand the degradation mechanisms of the drug substance to see if there is any potential impurity growth due to the excipients (and manufacturing process) chosen for drug product development. Compatibility studies between the drug substance and various excipients help in establishing an understanding of whether a particular combination of an

Table 6.3 Key considerations or design of solid oral dosage forms

	Purpose
Stability	There are numerous physical factors, such as temperature, moisture, light, etc., that influence the chemical degradation of pharmaceutical drugs. Therefore, stability testing is essential for the patient's safety. In order to maintain efficacy throughout the dosing regimen, the patient should receive a uniform dose of the drug throughout the product's shelf-life. In addition, although a drug may have been shown to be safe for use, this is not necessarily true of the decomposition product(s). It is, therefore, necessary to prevent (or minimize) the decomposition of the product. Instability is not limited to chemical changes only. Physical changes such as moisture content changes, changes in tablet hardness, color, friability, disintegration time, dissolution, etc., can also happen and need to be accordingly studied. It is typical for a dosage form to have an appropriate shelf life of at least 2–3 years
Bioavailability	Biopharmaceutics is defined as the study of factors influencing the rate and amount of drug that reaches the systemic circulation and the use of this information to optimize the therapeutic efficacy of the drug products. The process of movement of a drug from its site of administration to the systemic circulation is called absorption. Bioavailability is defined as the rate and extent (amount) of drug absorption. The concentration of the drug in plasma, hence the onset of action, and the intensity and duration of response depend upon the bioavailability of the drug from its dosage form
Manufacturability	Manufacturability is typically defined as the ease by which a formulation lends itself to processing. Clearly, formulation development and manufacturability are symbiotic and iterative processes. However, in the context of the long drug development cycle, it is essential to understand that the initial manufacturability that may have been gained in early-stage formulations may not present itself as the process is scaled up or equipment design changes occur
Acceptability	A product must be acceptable to those who handle or use it. The doctor, for instance, will note those features of a product that facilitate its administration to the patient. In this regard, solid oral dosages have a clear and acceptable precedence and, therefore, tend to be the favored route of administration for most patients. However, a formulator must still understand the nuances of acceptability in terms of dosage size, taste, odor, primary packaging, and regulatory compliance of ingredients
Business acuity	A formulator needs to realize that decisions made during the design phase of a product's development determine most of the costs of manufacturing that product. As the design and manufacturing processes become more complex and increase in scale, the formulator will increasingly be called upon to make decisions that involve a significant investment of resources in terms of time, people, and money. Each of these decisions cannot be made in isolation, and a balance must be struck each time to ensure that none of the other design rules of stability, bioavailability, manufacturability, and acceptability are violated. Poor product design without the foresight of scale-up, marketing, and global compliance may lead to wasting precious resources that could have been invested in the company's pipeline. An awareness of this concept is known as business acuity

excipient(s) with the drug substance will lead to any potential stability issues or not. Once the initial composition of the drug product is agreed upon, it is imperative to at least think about what kind of processing strategy will be employed. However, the formulation development and process selection are not mutually exclusive choices but an iterative exercise. For example, selecting the excipients could impact the manufacturing process and vice versa.

Most of the pharmaceutical materials are available as powders. Powders are multi-phasic materials. For example, powders exhibit mechanical behavior that apparently is similar to those exhibited by solids and/or fluids. However, there are certain attributes in the mechanical behavior of powders that cannot be completely explained either by solid mechanics or by fluid mechanics. All pharmaceutical process development efforts can be visualized as unit operations. These unit operations can range all the way from granulation, milling, and coating (Table 6.4). A detailed understanding of all the various unit operations can help give a formulator the tools to design the best process for their drug.

6.2.3 Challenges in Small Molecules Product Development

Development of small molecule-based drug products has been going on for some time now. However, a key question during drug development is whether a drug will be bioavailable after its administration. After all, good bioavailability facilitates formulation design and development, reduces inter-subject variability, and enhances dosing flexibility. Over the past two decades, there has been an increasing effort to bring new drug solubilization and delivery technologies earlier in the development cycle and enable successful new molecular entity progression to become viable clinical and commercial therapies. Advanced delivery technologies can also be applied in the early discovery space for the evaluation of critical "tool" molecules, such as known competitor compounds, as well as positive or negative control compounds in exploratory preclinical proof of concept or safety studies. The need for enabled drug delivery measures to support a wide variety of preclinical studies has

Table 6.4 Commonly used unit operations in solid oral dosage development

Unit operation	Examples
Particle size enlargement	Granulation, extrusion, spheronization
Particle size reduction	Delumping, screening, and milling
Drying	Convection, fluidized bed
Blending	Low and high shear blending
Filling	Encapsulation
Compression	Roller compaction, slugging, and tableting
Coatings	Particle coating and tablet coating

been greatly facilitated by the inclusion of formulation and pharmaceutical scientists in drug discovery teams. In developing the clinical formulations, it is extremely important for pharmaceutical scientists to have a comprehensive understanding of the planned clinical dose range, the dosing protocol (including in-clinic vs. take-home dosing and feeding state), the compound's absorption parameters, the desired PK profile, and any disease/physiological parameters that might impact absorption in the clinical subjects. It is also important to ensure that the physical and chemical stability profile of the compound in question is well understood and will not impact the delivery and bioavailability of the dose. Lastly, to the extent possible, clinical formulations should be developed with an appropriate line of sight for commercial development (Landis et al., 2018).

Another challenge encountered during product development is that of the potential for failures in clinical trials. For every 250 drugs that enter pre-clinical testing, only five will ever make it to human clinical trials, according to PhRMA. However, the trend in drug failures has changed during the past 20 years, as discussed in Sect. 1.3.3. The comparison of clinical attrition trends between 1991 and 2000 indicates that the principal component of drug failure in 1991 was the unacceptable pharmacokinetics profile in humans. Since then, the industry has adopted several preclinical screens to address permeability, metabolism, distribution, and excretion issues, including allometric scaling for the projection of human pharmacokinetics profile (as shown in Fig. 1.12). However, during human clinical trials, >30% of drugs entering Phase II studies fail to progress, and >58% of drugs go on to fail in Phase III (Van Norman, 2019). Analysis done by researchers on 87 failed Phase II drugs during 2008–2010 and on 83 failed Phase III drugs during 2007–2010 indicate insufficient efficacy to be the foremost reason for failure (as shown in Fig. 1.13).

Other researchers have also produced similar analyses. In a study conducted, researchers found that in 1991, the major causes of drug failure were poor PK properties (40%), lack of efficacy (30%), and toxicological issues (21%); by 2000, the complexion had changed considerably, with toxicity (32%) and efficacy issues (26%) being the major contributors, whereas PK issues accounted for just 8% of the failures. The combination of CMC issues, commercial reasons, and cost of goods together amounted to 5% of the failures in 1991, but by 2000, these factors accounted for 32%, with commercial reasons, the major contributor at 20%, reflecting an escalation in the competitive nature of the pharmaceutical business environment. By 2012, this scenario appeared to be little changed, with Phase I and II failures due to a lack of efficacy estimated at 56%, toxicity accounting for 28%, whereas strategic and business reasons amounted to 17% of the attrition, as seen in Table 6.5 (Meanwell, 2016). Discontinuations at the Phase III stage are the most problematic due to expenditures in both dollars and patient/investigator/industry resources.

Table 6.5 Sources of drug candidate attrition

Sources of drug candidate attrition	1991	2000	2011–12
Toxicity	21%	32%	28%
Lack of efficacy	30%	26%	56%
Pharmacokinetic/bioavailability	40%	8%	Not reported
Chemistry, manufacturing, and control issues	0%	4%	Not reported
Commercial	5%	20%	17%
Cost of goods	0%	8%	Not reported

Adapted with permission from Meanwell (2016). Copyright 2024 American Chemical Society.

6.3 Enhancing Value Proposition in Small Molecule Drug Development

The development of small-molecule drugs has historically been rooted in well-established manufacturing processes that have been around for a long time. Typical development operations include batch-process based manufacturing schemes. However, in August 2002, the FDA announced a significant new initiative, Pharmaceutical Current Good Manufacturing Practices (cGMPs) for the Twenty-First Century, to enhance and modernize the regulation of pharmaceutical manufacturing and product quality (as was discussed in Sect. 5.3.3). Since its inception in 2002, this proactive and encouraging approach from the FDA has yielded many benefits due to the active participation and input from experts in industry, academia, government, and consumer groups. Over the years, substantial progress has been made toward this vision, including process analytical technology (PAT), QbD, Emerging Technology, continuous manufacturing, and Six Sigma pharmaceutical quality (Yu et al., 2019). All these developments aim to enhance the value proposition of the pharmaceutical drug development process, with engineering principles and product design becoming the guiding principles of pharmaceutical development.

6.3.1 Enhanced Product Development through Modeling and Simulation

In healthcare-related academic and industrial research and development, modeling and simulation (M&S) (in silico) approaches are combined with other advanced scientific tools to support innovation for a better understanding of physiology, pathophysiology, complex diseases, and effects of medical interventions. The value of in silico methods in drug development and evaluation has been demonstrated repeatedly and convincingly (Musuamba et al., 2021). In the context of drug

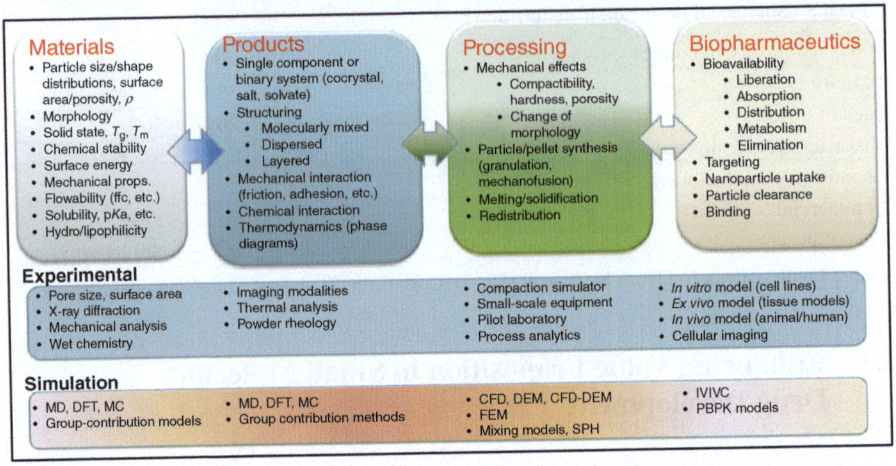

Fig. 6.6 Engineering view of pharmaceutical development. (Rantanen & Khinast, 2015)

product development, researchers have been increasingly employing in-silico methodologies to evaluate the impact of material properties on product performance and quality. In this new way of thinking, pharmaceutical ingredients, pharmaceutical products, related manufacturing processes, and biopharmaceutical properties are considered simultaneously and quantitatively. Simulation techniques include molecular dynamics (MD), density functional theory (DFT) computations, Monte Carlo methods, computational fluid dynamics (CFD), discrete element method (DEM), finite element method (FEM), smoothed particle hydrodynamics (SPH), in vitro-in vivo correlations (IVIVC), and physiologically based pharmacokinetic (PBPK) models (Rantanen & Khinast, 2015). This engineering viewpoint of pharmaceutical development is showcased in Fig. 6.6.

6.3.2 Embracing New Manufacturing Techniques and Approaches

In the last few years, there have been two important developments in the manufacturing of solid oral dosage products. These two developments are: (1) the wider usage of continuous manufacturing (CM) and (2) the advent of additive manufacturing (AM). CM is defined as a process in which the input material(s) are continuously fed into and transformed, and the processed output materials are continuously removed from the system. The system is defined as an integrated process that consists of two-or more-unit operations (Wahlich, 2021). A conceptual schematic of a CM-based end-to-end (E2E) process vs. a traditional batch process is shown in Fig. 6.7. Some benefits of adopting CM are given in Table 6.6. Since the pioneering work done at Rutgers University in the early 2000s, the pharmaceutical industry and

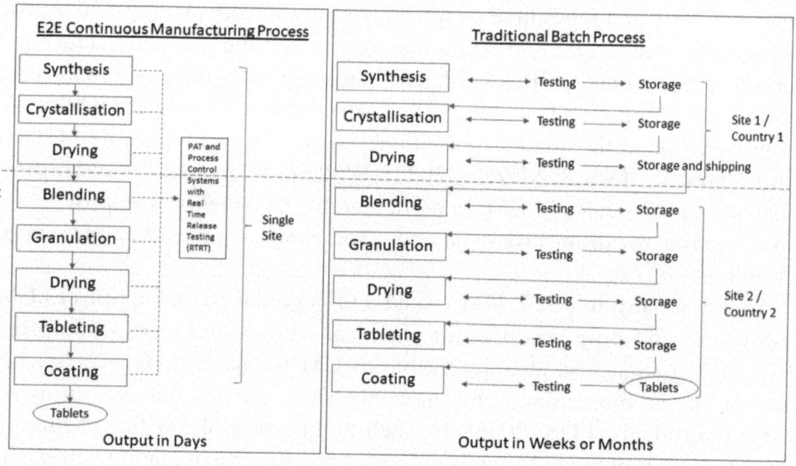

Fig. 6.7 Continuous manufacturing-based E2E process vs. traditional batch process. (Wahlich, 2021)

Table 6.6 Benefits of adopting CM (Wahlich, 2021)

Faster and leaner transition from development to commercial scale	Same equipment used during development, clinical supply manufacture and commercial production with batch sizes accommodated by changing run times Shorter cycle times More rapid development
Shorter supply chains	All operations conducted on one piece of equipment in one location without hold-ups and inter-site transfers between manufacturing stages Improved stability as no intermediate holding periods
Supply chain security	Enhanced domestic manufacture with reliance on high technology rather than low-cost labour Shorter manufacturing times meaning longer product shelf lives
Improved product quality	Domestic manufacture In-process monitoring, feedback and feed-forward controls. Maintenance of a state-of-control Reduced dependence on end-product testing Decreased regulatory oversight—frees resources for other higher-risk areas Fully aligned with principles of Quality by Design (QbD)
Cost benefits (after initial investment)	Lower production costs Improved utilisation of equipment Lower personnel requirements Smaller footprint
Supply chain responsiveness	Batch sizes tailored to requirements by adjusting run times Ability to respond rapidly to demands
Patient benefits	More suited to niche/personalised products Alternative manufacturing technologies (e.g., active pharmaceutical ingredient (API) printed or sprayed onto a dosage form) Lower risk of stock-outs
Societal benefits	Less environmental impact (less use of solvents, lower energy costs) Less waste/improved yields. Lowered risk that a whole batch will need to be rejected if it fails end-product testing Higher process intensification (less use of space, energy and raw materials) Improved safety (reduced handling and exposure to materials, easier cleaning) Source of high-tech jobs

its technology and ingredient suppliers have embraced a worldwide transformation from traditional, inefficient batch methods to continuous manufacturing, which is an emerging technology that has been shown to greatly reduce both the time and the cost of developing and manufacturing new medicines, while enabling significant improvements in the quality of the final product and the reliability of the manufacturing process (Oka & Muzzio, 2022). Worldwide, regulatory authorities have all been strong advocates of CM, in particular, for its benefits of improving the quality and control of drug products and producing a more flexible supply chain (Wahlich, 2021).

In that regard, in 2023, ICH issued a Q13 guidance that applies to CM of drug substances and drug products for chemical entities and therapeutic proteins. The ICH Q13 guidance is also applicable to CM for new products (e.g., new drugs, generic drugs, biosimilars) and the conversion of batch manufacturing to CM for existing products (FDA, 2023). In a self-audit report, the Office of Pharmaceutical Quality (OPQ) of the FDA analyzed time to approval and market entry, manufacturing process changes reported in Annual Reports, manufacturing-related post-approval application supplements, and pre-approval inspections. As per this report, at the start of 2022, there were six approved applications that utilized CM for their finished solid oral drug products. There were no substantial regulatory barriers identified for CM applications related to manufacturing process changes or pre-approval inspections. CM applicants had relatively shorter times to approval and market as compared to similar batch applications, based on the mean or median times to approval (8 or 3 months faster) and marketing (12 or 4 months faster) from submission, translating to an estimated \$171–537 M in early revenue benefit (Fisher et al., 2022).

Another major area of increasing research and industry interest is that of additive manufacturing (AM) which is also known as 3D printing (3DP). AM is a layer-by-layer fabrication method utilizing a printable material, or ink, to create a 3D object from a digital image developed via computer-aided design (CAD) (Abaci et al., 2021). The 3DP process was patented in the 1980s, but the utilization of this process has expanded substantially over the past decade, to which the pharmaceutical industry is a major contributor. The classification of various AM techniques is presented in Fig. 6.8. With increasing interest, researchers across the globe are striving for the fabrication of novel pharmaceutical dosage forms, especially tailored ones, which can cater to the specific needs of the patient. These dosage forms intend to cater to on–demand manufacturing, personalized medications, enhanced geometry, size, and dosage, and increased bioavailability of the medicinal active. With the emergence of precision medicine in healthcare, the inclusion of AM technologies is deemed imperative for the fabrication of oral dosage forms and polypills, which opens new horizons for the administration of drug combinations and formulations tailored to individual needs. Although the extensive commercialization and acceptance of AM techniques may disrupt the current healthcare supply chain, it has the potential to curtail the waste produced by expired and unused medications (Varghese et al., 2022). In 2015, the first FDA-approved drug made by a 3D printer was SPRITAM® (levetiracetam) tablets that are used as an adjunctive therapy in the

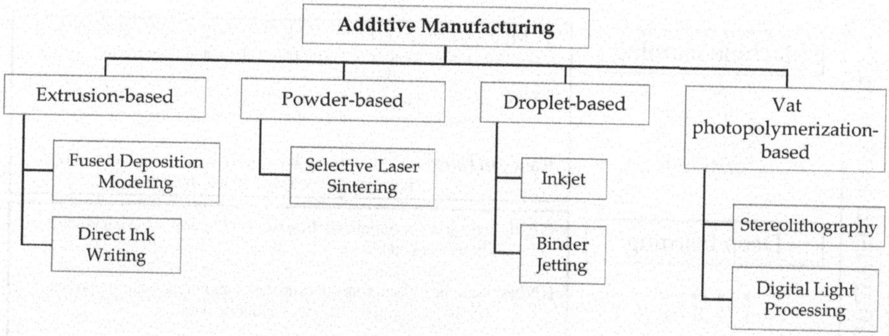

Fig. 6.8 Classification of AM techniques used in oral tablet fabrication. (Abaci et al., 2021)

treatment of partial onset seizures, myoclonic seizures, and primary generalized tonic-clonic seizures in people with epilepsy.

6.3.3 Use of Artificial Intelligence

Artificial intelligence (AI) has emerged as a powerful tool that harnesses anthropomorphic knowledge and provides expedited solutions to complex challenges. Remarkable advancements in AI technology and machine learning present a transformative opportunity in the drug discovery, formulation, and testing of pharmaceutical dosage forms. By utilizing AI algorithms that analyze extensive biological data, including genomics and proteomics, researchers can identify disease-associated targets and predict their interactions with potential drug candidates. This enables a more efficient and targeted approach to drug discovery, thereby increasing the likelihood of successful drug approvals. Furthermore, AI can contribute to reducing development costs by optimizing research and development processes. Machine learning algorithms assist in experimental design and can predict the pharmacokinetics and toxicity of drug candidates. This capability enables the prioritization and optimization of lead compounds, reducing the need for extensive and costly animal testing. Personalized medicine approaches can be facilitated through AI algorithms that analyze real-world patient data, leading to more effective treatment outcomes and improved patient adherence. In pharmaceutical product development, various AI models have been explored to enhance different aspects of the process. A list of commonly explored AI models in this domain is described in Fig. 6.9.

In the case of solid oral dosages, AI can help in the search for optimized formulation and the study of the desired attributes. AI can also be utilized in the context of systemic drug delivery to predict drug release profiles and dissolution profiles and explore the disintegration time for the effective selection of the best batch for further scale processing (Vora et al., 2023). Similarly, external information—including variables such as patient experience, market demand, supplier inventories, and

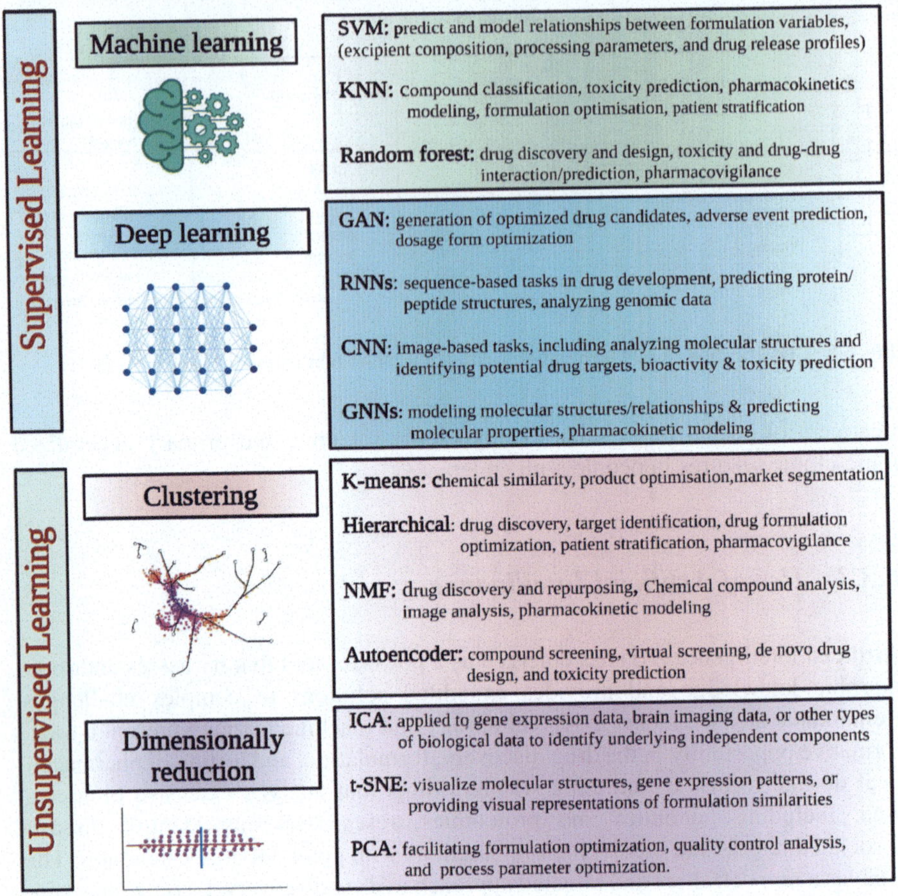

Fig. 6.9 Different supervised and unsupervised AI learning models/tools for pharmaceutical applications. (Vora et al., 2023)

public health emergencies—could fuse with internal information such as energy and resource management, modeling and simulation outcomes, and laboratory data. Integrating internal and external data sources enables unprecedented real-time responsiveness, monitoring, control, and prediction in pharmaceutical manufacturing. The result is a well-controlled, hyper-connected, digitized ecosystem and pharmaceutical value chain for the manufacturer, as shown in Fig. 6.10 (Arden et al., 2021). Such technologies are available today, and it behooves pharmaceutical scientists not only to strive to learn them but also to integrate this new knowledge into our traditional product development paradigm.

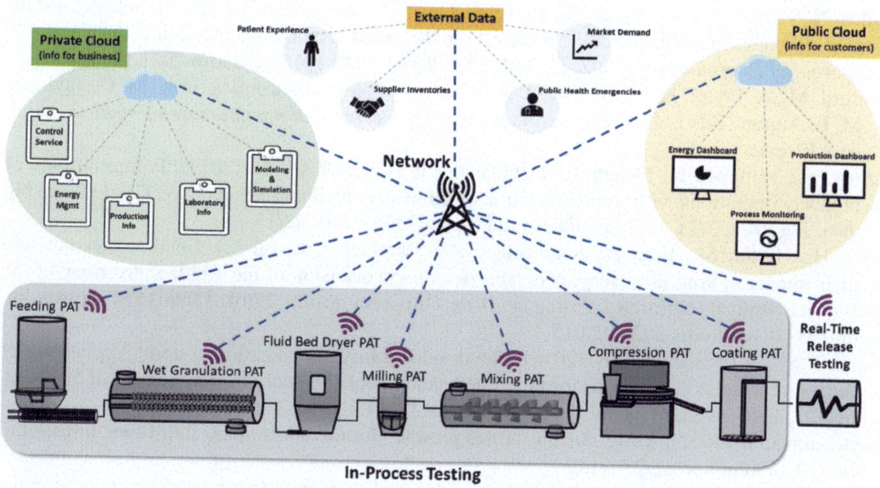

Fig. 6.10 A cyber-physical system (CPS) for pharmaceutical manufacturing in industry 4.0. (Arden et al., 2021) (Reprinted with permission from Elsevier)

6.4 Summary—What Else Is Possible—Part 1?

The future is always hard to predict. Not only does technology move at a breakneck speed, but also changes in society and socioeconomic patterns accelerate the innovation pathways. Pharmaceutical drug development is similarly poised for higher efficiencies through innovations in design, development, and manufacturing processes through the integration of modeling and simulation, artificial intelligence, and advanced manufacturing techniques. In all these innovations, though, it is critical that the savings are passed on to the stakeholders and that a concerted effort is made to lower the cost of goods. Given the dwindling physical resources of the planet, climate change, and associated economic boom-bust cycles coming in the future, such a cost-saving commitment is critical to help achieve global health equality and improve access to medications. ***Is making such a commitment not the right thing to do?***

References

Abaci, A., Gedeon, C., Kuna, A., & Guvendiren, M. (2021). Additive manufacturing of oral tablets: Technologies, materials and printed tablets. *Pharmaceutics, 13*(2). https://doi.org/10.3390/pharmaceutics13020156

Amidon, G. L., Lennernäs, H., Shah, V. P., & Crison, J. R. (1995). A theoretical basis for a biopharmaceutic drug classification: The correlation of in vitro drug product dissolution and in vivo bioavailability. *Pharmaceutical Research, 12*, 413–420. https://deepblue.lib.umich.edu/bitstream/handle/2027.42/41443/11095_2004_Article_306840.pdf?sequence=1

Arden, N. S., Fisher, A. C., Tyner, K., Yu, L. X., Lee, S. L., & Kopcha, M. (2021). Industry 4.0 for pharmaceutical manufacturing: Preparing for the smart factories of the future. *International Journal of Pharmaceutics, 602*, 120554. https://doi.org/10.1016/j.ijpharm.2021.120554

Ashford, M., & Taylor, K. M. G. (2022). Introduction to biopharmaceutics. In K. M. G. Taylor & M. E. Aulton (Eds.), *Aulton's pharmaceutics: The design and manufacture of medicines* (6th ed.). Elsevier.

Azmat, M., Kummer, S., Moura, L. T., Gennaro, F. D., & Moser, R. (2019). Future outlook of highway operations with implementation of innovative technologies like AV, CV, IoT and big data. *Logistics, 3*(2), 15. https://www.mdpi.com/2305-6290/3/2/15

Beck, H., Härter, M., Haß, B., Schmeck, C., & Baerfacker, L. (2022). Small molecules and their impact in drug discovery: A perspective on the occasion of the 125th anniversary of the Bayer Chemical Research Laboratory. *Drug Discovery Today, 27*(6), 1560–1574. https://doi.org/10.1016/j.drudis.2022.02.015

Butler, J. M., & Dressman, J. B. (2010). The developability classification system: Application of biopharmaceutics concepts to formulation development. *Journal of Pharmaceutical Sciences, 99*(12), 4940–4954. https://doi.org/10.1002/jps.22217

Doytchinova, I. (2022). Drug design—Past, present, future. *Molecules, 27*, 1496. https://doi.org/10.3390/molecules27051496

FDA. (2023). *Q13 continuous manufacturing of drug substances and drug products: Guidance for industry*. Retrieved from https://www.fda.gov/media/165775/download

Fisher, A. C., Liu, W., Schick, A., Ramanadham, M., Chatterjee, S., Brykman, R., Lee, S. L., Kozlowski, S., Boam, A. B., Tsinontides, S. C., & Kopcha, M. (2022). An audit of pharmaceutical continuous manufacturing regulatory submissions and outcomes in the US. *International Journal of Pharmaceutics, 622*, 121778. https://doi.org/10.1016/j.ijpharm.2022.121778

Göke, K., Lorenz, T., Repanas, A., Schneider, F., Steiner, D., Baumann, K., Bunjes, H., Dietzel, A., Finke, J. H., Glasmacher, B., & Kwade, A. (2018). Novel strategies for the formulation and processing of poorly water-soluble drugs. *European Journal of Pharmaceutics and Biopharmaceutics, 126*, 40–56. https://doi.org/10.1016/j.ejpb.2017.05.008

Landis, M. S., Bhattachar, S., Yazdanian, M., & Morrison, J. (2018). Commentary: Why pharmaceutical scientists in early drug discovery are critical for influencing the design and selection of optimal drug candidates. *AAPS PharmSciTech, 19*(1), 1–10. https://doi.org/10.1208/s12249-017-0849-3

Lay, M. G., & Benson, F. J. (2022). Road. In *Encyclopedia Britannica*.

LeDoux, M. (2016). *The history of compounds, extraction and tablet compression*. Retrieved January 31, 2024, from https://www.naturalproductsinsider.com/manufacturing/the-history-of-compounds-extraction-and-tablet-compression

Mahmood, I., Green, M. D., & Fisher, J. E. (2003). Selection of the first-time dose in humans: Comparison of different approaches based on interspecies scaling of clearance. *Journal of Clinical Pharmacology, 43*, 692–697. https://accp1.onlinelibrary.wiley.com/doi/abs/10.1177/0091270003254631?sid=nlm%3Apubmed

Mandal, S., Moudgil, M., & n., & Mandal, S. K. (2009). Rational drug design. *European Journal of Pharmacology, 625*(1), 90–100. https://doi.org/10.1016/j.ejphar.2009.06.065

Meanwell, N. A. (2016). Improving drug design: An update on recent applications of efficiency metrics, strategies for replacing problematic elements, and compounds in nontraditional drug space. *Chemical Research in Toxicology, 29*(4), 564–616. https://doi.org/10.1021/acs.chemrestox.6b00043

Mittal, B. (2016). *How to develop robust solid oral dosage forms: From conception to post-approval*. Elsevier/Academic Press.

Musuamba, F. T., Skottheim, R. I., Lesage, R., Russo, G., Bursi, R., Emili, L., Wangorsch, G., Manolis, E., Karlsson, K., Kulesza, A., Courcelles, E., Boissel, J., Rousseau, C., Voisin, E., Alessandrello, R., Curado, N., Dall'ara, E., Rodriguez, B., Pappalardo, F., & Geris, L. (2021). Scientific and regulatory evaluation of mechanistic in silico drug and disease models in drug

development: Building model credibility. *CPT: Pharmacometrics and Systems Pharmacology, 10*(8), 804–825. https://doi.org/10.1002/psp4.12669
Oka, S., & Muzzio, F. J. (2022). Chapter 1—Introduction. In F. J. Muzzio & S. Oka (Eds.), *How to design and implement powder-to-tablet continuous manufacturing systems* (pp. 1–8). Academic. https://doi.org/10.1016/B978-0-12-813479-5.00008-2
Rantanen, J., & Khinast, J. (2015). The future of pharmaceutical manufacturing sciences. *Journal of Pharmaceutical Sciences, 104*(11), 3612–3638. https://doi.org/10.1002/jps.24594
Roser, M. (2023). *Technology over the long run: Zoom out to see how dramatically the world can change within a lifetime*. https://ourworldindata.org/technology-long-run
Swann, J. P. (1995). The evolution of the American pharmaceutical industry. *Pharmacy in History, 37*(2), 76–86.
Uhlemann, J., Diedam, H., Hoheisel, W., Schikarski, T., & Peukert, W. (2021). Modeling and simulation of process technology for Nanoparticulate drug formulations—A particle technology perspective. *Pharmaceutics, 13*(1), 22. https://www.mdpi.com/1999-4923/13/1/22
Van Norman, G. A. (2019). Phase II trials in drug development and adaptive trial design. *JACC Basic to Translational Science, 4*(3), 428–437. https://doi.org/10.1016/j.jacbts.2019.02.005
Varghese, R., Salvi, S., Sood, P., Karsiya, J., & Kumar, D. (2022). Recent advancements in additive manufacturing techniques employed in the pharmaceutical industry: A bird's eye view. *Annals of 3D Printed Medicine, 8*, 100081. https://doi.org/10.1016/j.stlm.2022.100081
Vora, L. K., Gholap, A. D., Jetha, K., Thakur, R. R. S., Solanki, H. K., & Chavda, V. P. (2023). Artificial intelligence in pharmaceutical technology and drug delivery design. *Pharmaceutics, 15*(7). https://doi.org/10.3390/pharmaceutics15071916
Wahlich, J. (2021). Review: Continuous manufacturing of small molecule solid oral dosage forms. *Pharmaceutics, 13*(8). https://doi.org/10.3390/pharmaceutics13081311
Yu, L. X., Raw, A., Wu, L., Capacci-Daniel, C., Zhang, Y., & Rosencrance, S. (2019). FDA's new pharmaceutical quality initiative: Knowledge-aided assessment & structured applications. *International Journal of Pharmaceutics: X, 1*, 100010. https://doi.org/10.1016/j.ijpx.2019.100010

Chapter 7
Enhancing the Value Proposition of Biologics

Few scientists acquainted with the chemistry of biological systems at the molecular level can avoid being inspired

Donald Cram

Contents

Abstract Biology is amazing. The modern biotechnology industry has produced numerous cures for humanity's persistent diseases. However, biologics are expensive. In 2017, only 2% of US prescriptions comprised biologics, yet this small percentage accounted for 37% of net drug spending. Also, since 2014, 93% of net drug spending at a global level emanated from biologics. It is estimated that, on average, the daily dose of a biologic costs 22 times more than that of a small molecule. Given the complexity of manufacturing, the high costs of biologics are hardly a surprise. There are, however, some favorable developments. Researchers are integrating continuous manufacturing in their plants to improve scalability and facilitate time to market while lowering capital and operating costs and enhancing quality. Similarly, the advent of new process analytical tools to improve process robustness, accelerate

B. Mittal, *Sustainable Global Health Systems and Pharmaceutical Development*, AAPS Advances in the Pharmaceutical Sciences Series 60,
https://doi.org/10.1007/978-3-031-50415-0_7

scale-up to commercial production, and drive more efficient use of resources is on the rise. Manufacturers are also using single-use systems to increase flexibility and reduce production lead times while lowering capital investment and energy requirements. While all this is ongoing, the development of biosimilars is on the rise, which has the potential to further lower the costs of medicines.

Keywords Biologics drug development · Upstream processing · Downstream processing · Digital manufacturing · Single-use systems · Biosimilars

7.1 Introduction

We all have learned biology and the evolution of life in our early education. However, just because something is inherently familiar does not mean that we cannot pause to appreciate the advancement of humankind's knowledge and exploration with child-like wonder! Cell, in biology, is the basic membrane-bound unit that contains the fundamental molecules of life and of which all living things are composed. A single cell is often a complete organism in itself, such as a bacterium or yeast. Other cells acquire specialized functions as they mature. These cells cooperate with other specialized cells and become the building blocks of large multicellular organisms, such as humans and other animals. Human cells are only about 20 μm across, and each human is composed of more than 30,000,000,000,000 cells (Alberts et al., 2024).

The cell was first introduced as a material concept describing microscale chambers by Robert Hooke in 1665 (Li et al., 2021). A cell is enclosed by a plasma membrane, which forms a selective barrier that allows nutrients to enter and waste products to leave. The interior of the cell is organized into many specialized compartments, or organelles, each surrounded by a separate membrane. One major organelle, the nucleus, contains the genetic information necessary for cell growth and reproduction. Each cell contains only one nucleus, whereas other types of organelles are present in multiple copies in the cellular contents, or cytoplasm. Organelles include mitochondria, which are responsible for the energy transactions necessary for cell survival; lysosomes, which digest unwanted materials within the cell; and the endoplasmic reticulum and the Golgi apparatus, which play important roles in the internal organization of the cell by synthesizing selected molecules and then processing, sorting, and directing them to their proper locations (Fig. 7.1) (Alberts et al., 2024). After several hundreds of years of rapid development of biochemistry and molecular biology, researchers have identified many essential biomacromolecules contained inside cells. In such a tiny cell structure, there are as many as 42 million proteins, 3.6 million messenger RNAs, and 20–30 thousand genes (Li et al., 2021). It is a miracle of science that scientists have been able to harness the power of the cell through biotechnology.

The modern biotechnology industry emerged in the 1970s, based largely on a new recombinant DNA technique published in 1973 by Stanley Cohen of Stanford University and Herbert Boyer of the University of California. Recombinant DNA

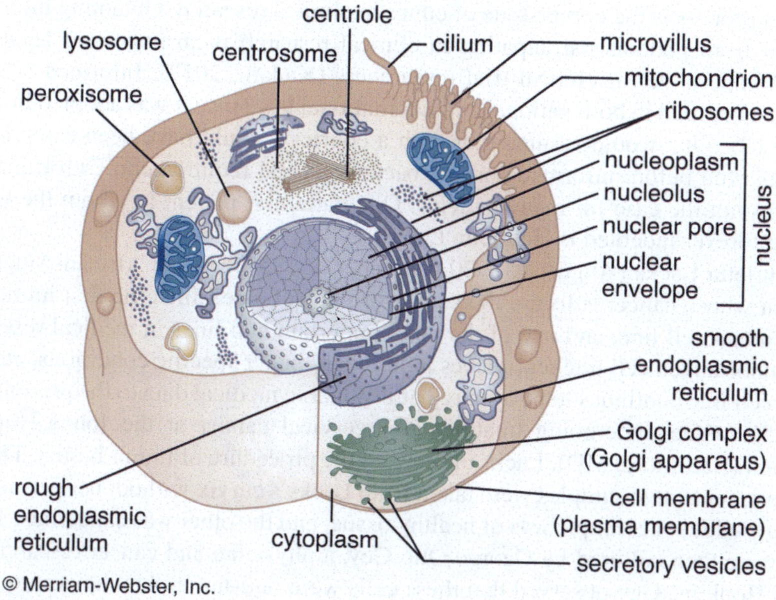

Fig. 7.1 Principal structures of an animal cell. (Alberts et al., 2024)

and subsequent discoveries have enhanced and accelerated our ability to develop practical biotechnology products to help us live longer and healthier lives, have a more abundant and sustainable food supply, use cleaner and more efficient processes for industrial manufacturing, and reduce our greenhouse gas footprint. By the beginning of the twentieth century, vaccines existed for rabies, diphtheria, typhoid fever, and the plague. By the 1990s, smallpox and polio had been eradicated or nearly eradicated worldwide, preventing millions of deaths. Worldwide, 2.5 million child deaths are prevented each year by immunization (Greenwood, 2014). In the past 20 years alone, the biopharmaceutical industry has added to our arsenal new vaccines for shingles, pneumonia, human papilloma virus (which causes a common form of cervical cancer), rotavirus (an infection fatal to many infants), and now COVID-19. One particular study material that helped in the development of drugs for numerous ailments was the HeLa immortal cell line.

7.1.1 *Philosophical Digression—HeLa—A Gift for Humanity*

The concept of "Informed Consent" emphasizes a process where the clinical research participant must receive and comprehend information appropriately to make an autonomous decision. An informed consent process can be termed as complete, valid, and meaningful if all four criteria of information disclosure, competence, comprehension, and voluntariness are effectively satisfied. The informed consent process is the cornerstone of ethics in clinical research. Obtaining informed consent from patients participating in clinical research is an important legal and ethical imperative for clinical trial researchers (Kadam, 2017). Informed consent has been codified in both national and international law since it was adopted in 1966 by the UN. One would assume that such a courtesy would have been extended to patients even before informed consent became a legal requirement. Unfortunately, that was not the case for many patients. One particular patient to whom the world will be forever indebted is Henrietta Lacks.

Henrietta Lacks (August 1, 1920—October 4, 1951) was an African-American woman whose cancer cells are the source of the HeLa cell line, the first immortalized human cell line, and one of the most important cell lines in medical research. An immortalized cell line reproduces indefinitely under specific conditions, and the HeLa cell line continues to be a source of invaluable medical data to the present day. In 1951, while undergoing treatment for cervical cancer at the Johns Hopkins Hospital (Baltimore, MD), Lacks underwent the procedure of tumor biopsy. During her treatments, two samples were taken from Lacks's cervix without her permission or knowledge; one sample was of healthy tissue, and the other was cancerous. These cells were then cultured by George Otto Gey, a physician and cancer researcher at Johns Hopkins. Gey observed that these cells were unusual in that they reproduced at a very high rate and could be kept alive long enough to allow more in-depth examination. Until then, cells cultured for laboratory studies survived for only a few days at most, which was not long enough to perform a variety of different tests on the same sample. Lacks's cells were the first to be observed that could be divided multiple times without dying, which is why they became known as "immortal". From the cells of the cancerous sample, Gey created the immortal cell line known as HeLa, which is still used for contemporary biomedical research (Various, 2024).

As was then the practice, no consent was required to culture the cells obtained from Lacks's treatment. Lacks passed away on October 4, 1951, from the cancer that had metastasized throughout her entire body. However, her memory lived on with the HeLa immortal cell line. Due to their ability to reproduce rapidly, HeLa cells became ubiquitous study material, contributing to the development of drugs for numerous ailments, including polio, Parkinson's disease, and leukemia (Etheredge, 2024). There is no way of knowing how many of Henrietta's cells are alive today. As per Rebecca Skloot, the author of a best-selling book on Henrietta Lacks' legacy, one scientist estimates that if you could pile all HeLa cells ever grown onto a scale, they would weigh more than 50 million metric tons (Skloot, 2010). In spite of this, until the 1970s, Henrietta's role was unknown even to her family. In the twenty-first century, Henrietta's case was an important component in the debate surrounding informed consent from patients for the extraction and use of cells in research (Etheredge, 2024).

Thankfully, recognition for Henrietta Lacks' contribution to scientific research was belated and acknowledged. In 2010, the Johns Hopkins Institute for Clinical and Translational Research established the annual Henrietta Lacks Memorial Lecture Series to honor Henrietta Lacks and the global impact of HeLa cells on medicine and research. Several other organizations followed suit, and on October

13, 2021, the WHO presented the Director General Award to Lawrence Lacks, the son of Henrietta Lacks, in recognition of her unknowing contribution to science and medicine (Various, 2024).

7.1.2 *Opportunity for Pharmaceutical Scientists: Acknowledge and Appreciate the Patient Interface*

Patients are at the center of healthcare. Not only are they the consumers of the products that the pharmaceutical industry produces, but their input and their cooperation are fundamental to the advancement of medical sciences. In 1988, the Picker Commonwealth Program for Patient-Centered Care's research program, based at Boston's Beth Israel Deaconess Hospital, conducted national qualitative research to identify the aspects of care most valued by patients to guide the development of patient experience surveys that could be used to evaluate care. Patient-centeredness was defined as "Health care that establishes a partnership among practitioners, patients, and their families to ensure that decisions respect patients' wants, needs, and preferences and that patients have the education and support they need to make decisions and participate in their own care." The Picker/Commonwealth work and later the Picker Institute's research confirmed that patients are often the only ones who can evaluate many aspects of care and its quality (Edgman-Levitan & Schoenbaum, 2021).

It does not matter to the patient what providers think they have communicated or documented in the chart. If the patient does not understand or remember it, the quality of care suffers. The Picker approach to measuring the patient's experience of care, as opposed to their satisfaction with care, has continued through the work of the US Consumer Assessment of Healthcare Providers and Systems (CAHPS) Consortium, funded by the Agency for Healthcare Research and Quality (AHRQ) [1] That work has developed and implemented national standardized patient experience surveys in the US that have been in use since the mid-1990s. In 2001, the United States Institute of Medicine, now the National Academy of Medicine, issued its landmark report, "Crossing the Quality Chasm". This report makes an urgent call for fundamental change to close the quality gap. It recommends a sweeping redesign of the American health care system and provides overarching principles for specific direction for policymakers, health care leaders, clinicians, regulators, purchasers, and others (Institute_of_Medicine, 2001). It called for a transformation of the US healthcare system to be guided by six aims: Safety, Effectiveness, Efficiency, Patient-Centeredness, Timeliness, and Equity. The report adopted Picker's definition of patient-centered care. Since its publication, many healthcare organizations,

[1] AHRQ is the lead Federal agency charged with improving the safety and quality of America's health care system. AHRQ develops the knowledge, tools, and data needed to improve the health care system and help Americans, health care professionals, and policymakers make informed health decisions. AHRQ works within the Department of Health and Human Services (HHS).

both public and private, have worked to improve these aims with varying success (Edgman-Levitan & Schoenbaum, 2021).

In the context of pharmaceutical product development, patient-centricity involves the patient in this scientific process. In essence, this is an innovative mindset whereby the pharmaceutical industry is challenged to engage and collaborate with patients when deciding the best course of action. FDA also encourages patient-focused drug development (PFDD) as part of the requirements of the twenty-first Century Cures Act and as part of commitments made by the FDA under the sixth authorization of the Prescription Drug User Fee Act (PDUFA). FDA defines PFDD as a systematic approach to help ensure that patients' experiences, perspectives, needs, and priorities are captured and meaningfully incorporated into the development and evaluation of medical products throughout the medical product life cycle (FDA, 2018a).

In Sect. 5.3.4, the patient-centric approach to drug product design was introduced. Patient-centric product design is the process of identifying the comprehensive needs of individuals or the target patient population and utilizing the identified needs to design pharmaceutical drug products that provide the best overall benefit to the risk profile for that target patient population over the intended duration of treatment. Numerous product-related and patient-related characteristics that can help in developing a patient-centric pharmaceutical drug product were discussed with the aim of improving medication adherence. However, an important element of patient-focused drug development is to be also aware of the costs of the medication that the patients will ultimately have to pay (whether they have insurance or not).

There is a vast difference in the commercialization outlook of the two pharmaceutical product categories: small molecules and biologics. In 2017, only 2% of US prescriptions comprised biologics, yet this small percentage accounted for 37% of net drug spending. Also, since 2014, 93% of net drug spending at a global level emanated from biologics. Nevertheless, in today's drug market, small molecules constitute as much as 90% of global sales. Nevertheless, in the United States and other countries, the rate of purchase of biologics by those who can afford them is increasing alarmingly. It is estimated that, on average, the daily dose of a biologic costs 22 times more than that of a small molecule (Makurvet, 2021). ***What makes biologics so expensive, and what can be done to lower their costs and improve their efficiency?***

7.2 Biologics Product Development

Biologicals or biologics, including those made by biotechnology, are a special category of "drugs" or medicines. They differ from pharmaceutical drugs derived by chemical means in that they are derived biologically by the growth of microorganisms or cells (often mammalian, including human cells). Biologicals usually consist of proteins, carbohydrates, nucleic acids, cells or tissues for transplantation, or a complex composite of these substances. Examples include hormones, vaccines,

Table 7.1 Selected categories of biological agent structure

Biologic agent structure	Details
Hormone	A substance, usually a peptide or steroid, is produced by one tissue and conveyed by the bloodstream to another to affect physiological activity, such as growth or metabolism
Interferons	Proteins that are normally produced by cells in response to viral infection and other stimuli
Interleukins	A large group of cytokine proteins. Most are involved in directing other immune cells to divide and differentiate
Growth factor	A substance such as a vitamin B12 or an interleukin that promotes growth, especially cellular growth
Monoclonal antibodies (mAbs)	A single species of immunoglobulin molecules is produced by culturing a single clone of a hybridoma cell. MAbs recognize only one chemical structure, i.e., they are directed against a single epitope of the antigenic substance used to raise the antibody
Polypeptides	Peptides contain ten or more amino acids. Typically, a peptide consists of fewer than 50 amino acids, while a protein has more than 50 amino acids
Proteins	Naturally occurring and synthetic polypeptides have molecular weights greater than about 10,000 (the limit is not precise)
Vaccine	An agent containing antigens produced from killed, attenuated, or live pathogenic microorganisms, synthetic peptides, or recombinant organisms. They are used for stimulating the immune system of the recipient to produce specific antibodies providing active immunity and/or passive immunity in the progeny

Adapted from Morrow and Felcone (2004)

blood products, allergenics, monoclonal antibodies, recombinant therapeutic proteins, gene and cellular therapies, growth factors, cytokines, and insulin, among others (Makurvet, 2021). A few of the biological agent structures are discussed in Table 7.1.

Most biologicals are large, complex molecules that biologics are typically greater than 1000 Daltons (Da) in size compared to small molecules that are relatively smaller, usually between 100 and 1000 Da (Sheets, 2018). This difference in molecular weight is demonstrated in Fig. 7.2, where a large molecule (monoclonal antibody) is compared with a small molecule (aspirin). A single monoclonal antibody weighs more than 800 times what an aspirin molecule weighs (FDA, 2023a). Examples of biologic products with their corresponding molecular weights include filgrastim (Neupogen®) (18,800 Da), interferon alfa-2b (Roferon-A®) (19,625 Da), (epoetin alfa (Epogen®) (30,400 Da), etanercept (Enbrel®) (75,000 Da), daclizumab (Zenapax®) (144,000 Da), rituximab (Rituxan® and MabThera®) (145,000 Da), and Factor VIII (an essential blood-clotting protein) (264,000 Da) (Tariman, 2018).

Apart from molecular weight, there are significant differences in the way biological products are manufactured when compared to small molecules-based products. Biologics are notoriously sensitive to a given manufacturing process and the starting materials. Biological products, including those manufactured by

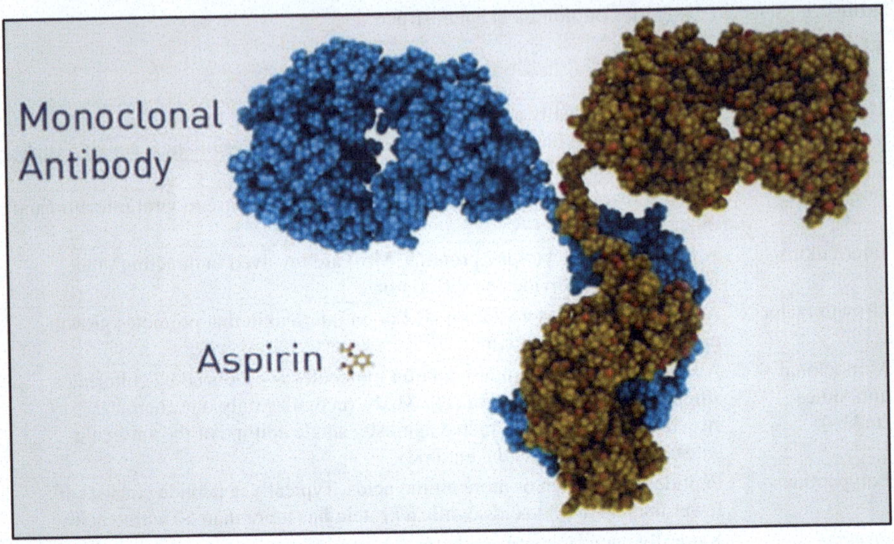

Fig. 7.2 Structural comparison of a monoclonal antibody vs. aspirin (small molecule). (FDA, 2023a)

biotechnology, tend to be heat-sensitive and susceptible to microbial contamination. Therefore, it is necessary to use aseptic principles from the initial manufacturing steps. The structural complexity of biologics makes characterization difficult; hence, clinical effects are hardly predictable in patients (Makurvet, 2021). Another difference is the regulatory path followed for approval. In the United States, small molecule pharmaceuticals are regulated and licensed for marketing under the Food Drug and Cosmetic Act (FDCA), whereas, biologic products are regulated and approved under the Public Health Services Act (PHSA).

7.2.1 *Statistics of Biologics*

Biological products often represent cutting-edge biomedical research and, in time, may offer the most effective means to treat a variety of medical illnesses and conditions that presently have no other treatments available (FDA, 2018b). The first monoclonal antibody (mAb) was generated in mice in 1975 when Georges J.F. Köhler and César Milstein developed a hybridoma method for the production of monoclonal antibodies. They were awarded the Nobel Prize in Physiology or Medicine in 1984 jointly with Niels K. Jerne for theories concerning the specificity in the development and control of the immune system and the discovery of the principle for the production of monoclonal antibodies. The first monoclonal antibody that was fully licensed in 1986 was Orthoclone OKT3 (muromonab-CD3), which was approved for use in preventing kidney transplant rejection. Today, with major

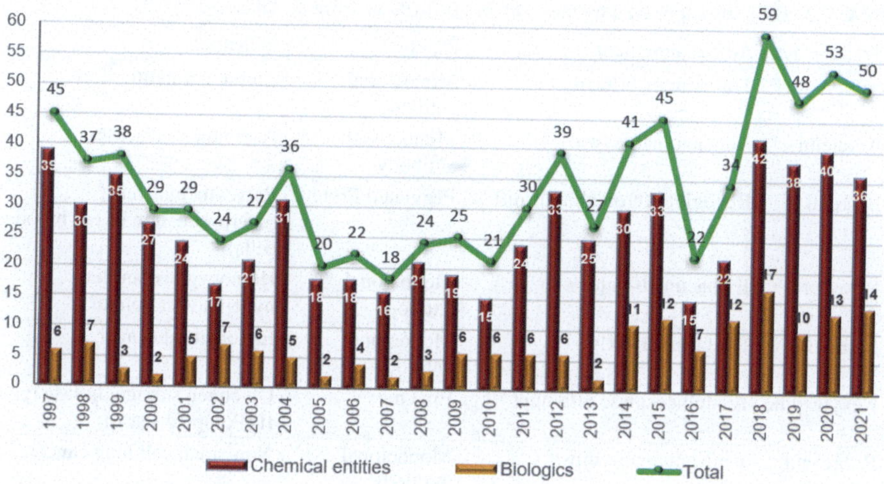

Fig. 7.3 New chemical entities and biologics approved by FDA (1997–2021). (de la Torre & Albericio, 2022)

rapid advancements in genetic sequencing and the translation of basic medical sciences research into clinical practice, humanized monoclonal antibodies are now the fastest-growing group of biotechnology-derived molecules in clinical trials. Monoclonal antibodies are currently approved by the FDA for human use to treat various diseases and conditions, including cancer, chronic inflammatory diseases, transplantation, infectious diseases, and cardiovascular diseases (Liu, 2014).

In the last 25 years, the share of biologics approved by the FDA has steadily risen, as shown in Fig. 7.3. Biologics continue to account for slightly more than 25% of all drugs accepted by the FDA. In 2021, 14 biologics were approved, of which eight were monoclonal antibodies, two antibody drug conjugate (ADC), two enzymes, and two pegylated biomolecules, namely a hormone and a protein (Table 7.2). Therefore, mAbs continue to be the most prevalent drug of all types of biologics (de la Torre & Albericio, 2022).

An important and persistent aspect of biologics drug products is their high prices. According to an article published by Forbes, in 2017, only 2% of US prescriptions comprised biologics, yet this small percentage accounted for 37% of net drug spending (Fig. 7.4). Also, since 2014, 93% of net drug spending at a global level has emanated from biologics (Roy, 2019). One of the reasons that the spending on biologics is so high is that biologics aim to treat hitherto untreated conditions that have irrevocable consequences on morbidity and mortality. Development timelines stretch into decades, contributing in no small measure to sunk costs that cannot be recovered in ways other than through higher prices of commercialized products. Not inconsequentially, several biologics are novel, first-in-class entities that enjoy virtual monopolies with the ability to command high prices. In-market, monoclonal antibodies typically cost upward of $15,000 a year. Biologic treatments for cancer or rare, life-threatening diseases can cost anywhere from $50,000 to $400,000 per

Table 7.2 Biologics approved by the FDA in 2021 (de la Torre & Albericio, 2022)

Trade name	Active ingredient	Class	Indication
Adbry	Tralokinumab-ldrm	Monoclonal antibody	Atopic dermatitis
Aduhelm	Aducanumab-avwa	Monoclonal antibody	Alzheimer's disease
Besremi	Ropeginterferon alfa-2b-njft	Pegylated Protein	Polycythemia vera (overproduction of red blood cells)
Evkeeza	Evinacumab-dgnb	Monoclonal antibody	Homozygous familial hypercholesterolemia
Jemperli	Dostarlimab-gxly	Monoclonal antibody	Endometrial cancer
Nexviazyme	Avalglucosidase alfa-ngpt	Enzyme	Glycogen storage disease type II (Pompe disease)
Rybrevant	Amivantamab-vmjw	Monoclonal antibody	Non-small cell lung cancer
Rylaze	Asparaginase erwinia chrysanthemi (recombinant)-rywn	Enzyme	Acute lymphoblastic leukemia and lymphoma
Saphnelo	Anifrolumab-fnia	Monoclonal antibody	Systemic lupus erythematosus
Skytrofa	Lonapegsomatropin-tcgd	Pegylated hormone	Growth hormone deficiency
Tezspire	Tezepelumab	Monoclonal antibody	Asthma as maintenance therapy
Tivdak	Tisotumab vedotin-tftv	Antibody-drug conjugate	Cervical cancer
Vyvgart	Efgartigimod alfa-fcab	Monoclonal antibody (fragment)	Myasthenia gravis
Zynlonta	Loncastuximab tesirine-lpyl	Antibody-drug conjugate	Relapsed or refractory large B-cell lymphoma

year (Rao, 2011). It is, therefore, critical to understand the biologics product development process to appreciate the complexity of producing biologics commercially.

7.2.2 *Typical Biological Product Development*

Biologics manufacturing (or biomanufacturing) is a complex process that produces a product largely derived from discoveries using recombinant DNA technology to develop processes and analytics to manufacture biotherapeutic products. These recombinant products were developed from several platforms, such as whole multicellular systems encompassing transgenic plants, animals, unicellular microbes (bacteria and yeast), and insect and mammalian cell cultures. The biological

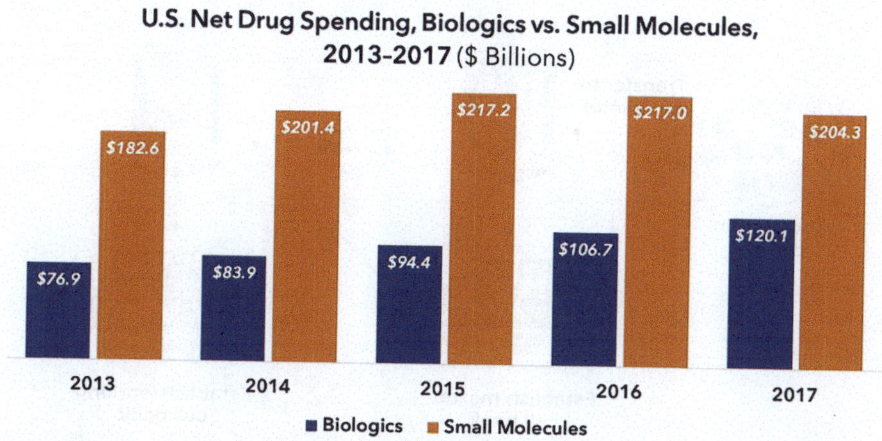

Fig. 7.4 US net drug spending on biologics and small molecules (2013–2017). (Roy, 2019)

product to be developed and manufactured starts with the discovery and proof-of-concept studies characterizing the product as well as defining the actual molecule and its mode of action. The study and knowledge of the molecule's characteristics will be important in designing cell culture and bioreactor parameters in order to scale up and produce a sufficient yield of product to satisfy various early experiments such as in vitro and in vivo (small animal) studies, process development, formulation, and analytical development studies. These studies help to characterize the biological molecule's phenotypic, genotypic, and biochemical profiles (Conner et al., 2020).

The manufacturing of biologics begins with the expression in biologic, cell-based systems of proteins that then require recovery and purification, often by multiple chromatographic techniques. The general overall bioproduction process is executed in a series of unit operations (Shire, 2009). A brief overview of these unit operations is provided below and illustrated in Figs. 7.5 and 7.6:

- ***Step 1 (Establishing the Cell Bank):*** A cell line that produces the monoclonal antibody is created using genetic engineering techniques. The two most commonly used production cells are E. coli bacterial cells and Chinese Hamster Ovary cells, or CHO cells. Once a manufacturer successfully manipulates a cell to produce said protein, the cells multiply. These genetically identical cells are known as the production cell line (Burke, 2023).
- ***Step 2 (Creation of Master Cell Bank):*** As the characteristics of a cell line affect the characteristics of the biologic it produces, it is essential for consistent production that, once selected for development, the cell line will never change. In practice, this is done by the creation of a master cell bank (MCB), where aliquots of the selected cell line are frozen at −180 °C and serve as the permanent reposi-

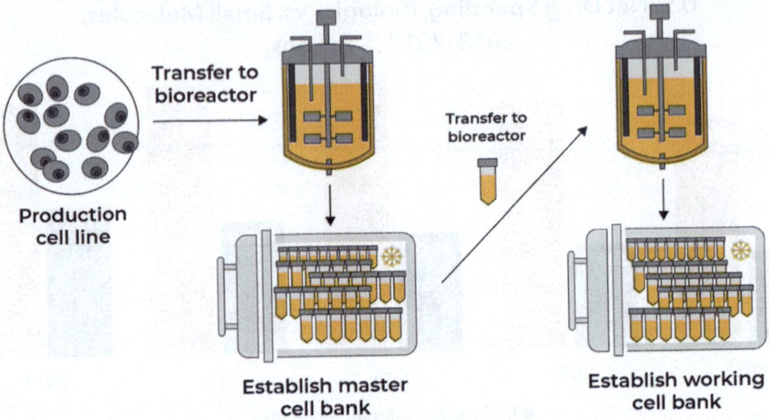

Fig. 7.5 Cell bank production. (Burke, 2023)

tory for future use. The MCB is kept at multiple locations for continuity of supply (Melsheimer et al., 2018).

- ***Step 3 (Creation of Working Cell Bank):*** Given the finite number of MCB aliquots, they must be used selectively. For this reason, a working cell bank (WCB) is created for routine production purposes. To create a WCB, an aliquot from the MCB is taken, grown in cell culture, re-aliquoted, and re-frozen (Fig. 7.5). An aliquot from the WCB is put into production for each batch of the product. When the WCB is depleted, a replacement WCB batch is prepared from the MCB (Melsheimer et al., 2018).
- ***Step 4 (Manufacturing of Product):*** The manufacturing of the final product from the WCB is itself a multistep process that requires meticulous planning and design of the facility, the utility/engineering systems, the equipment, identification, and validation of the conditions under which the cell line and the biologic are best maintained and processed (Fig. 7.6). It also requires the identification of the parameters that are key to measuring process performance and product quality attributes, which are key to safety, efficacy, and consistency. Once finalized, the manufacturing facility, the process, the identified parameters, and their specifications must be approved by regulators as a condition of marketing approval and are then locked into place. Commercial products and products used for subsequent clinical development are sourced from this approved process (Melsheimer et al., 2018).

Chemistry, Manufacturing, and Control (CMC) processes require that the development of biological drugs need to be conducted according to the existing regulatory guidelines and standards. Numerous ICH documents have been put in place to address quality and CMC processes involved in the characterization and

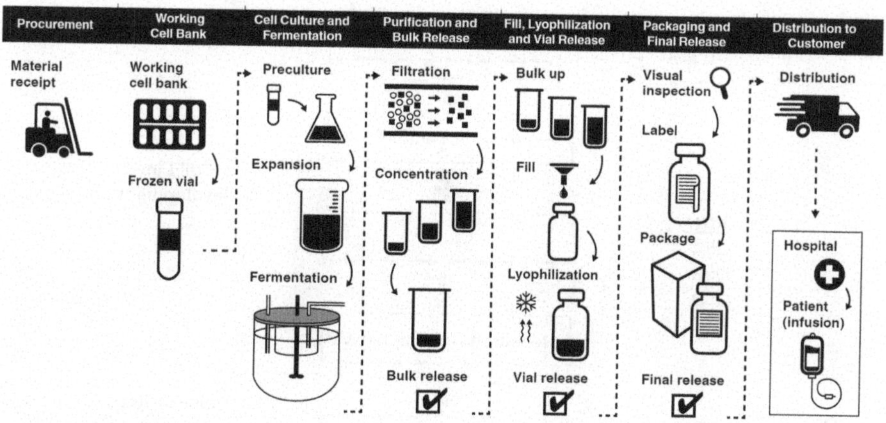

Fig. 7.6 Overview of biological product's manufacturing process. (Melsheimer et al., 2018)

manufacturing of biopharmaceuticals, reflecting both the complexity of these products and the high regulatory scrutiny surrounding their quality, as shown in Fig. 7.7 (Conner et al., 2020).

7.2.3 Challenges in Biological Product Development

An important challenge in the pharmaceutical development of a biologic is the optimization of safety and efficacy while ensuring the ability to manufacture the drug while maintaining quality and stability (Shire, 2009). As can be expected, quality control at each phase of manufacturing is paramount as the biologic is subjected to different stresses and conditions that may comprise the quality and stability. The manufacturer is responsible for establishing a rigorous and reliable quality system that ensures strict control over the manufacturing process and product quality attributes and fully adheres to GMPs specified by regulatory authorities. The entire production process is subject to testing at each step of the way, including quality control of raw materials, monitoring of equipment (operating parameters), the WCB aliquot (e.g., productivity, viability), the antibody (e.g., molecular weight, protein structure) and general product quality (e.g., chemical characteristics, potency, purity, sterility) (Melsheimer et al., 2018).

The nature of biological products, including the inherent variations that can result from the manufacturing process, can present challenges in characterizing and manufacturing these products that often do not exist in the development of small-molecule drugs. Slight differences between manufactured lots of the same biological product (i.e., acceptable within-product variations) are normal and expected within the manufacturing process. As part of its review, the FDA assesses the manufacturing process and the manufacturer's strategy to control within-product variations. These control strategies are put in place to help ensure that manufacturers produce biological products with consistent clinical performance (FDA, 2023a).

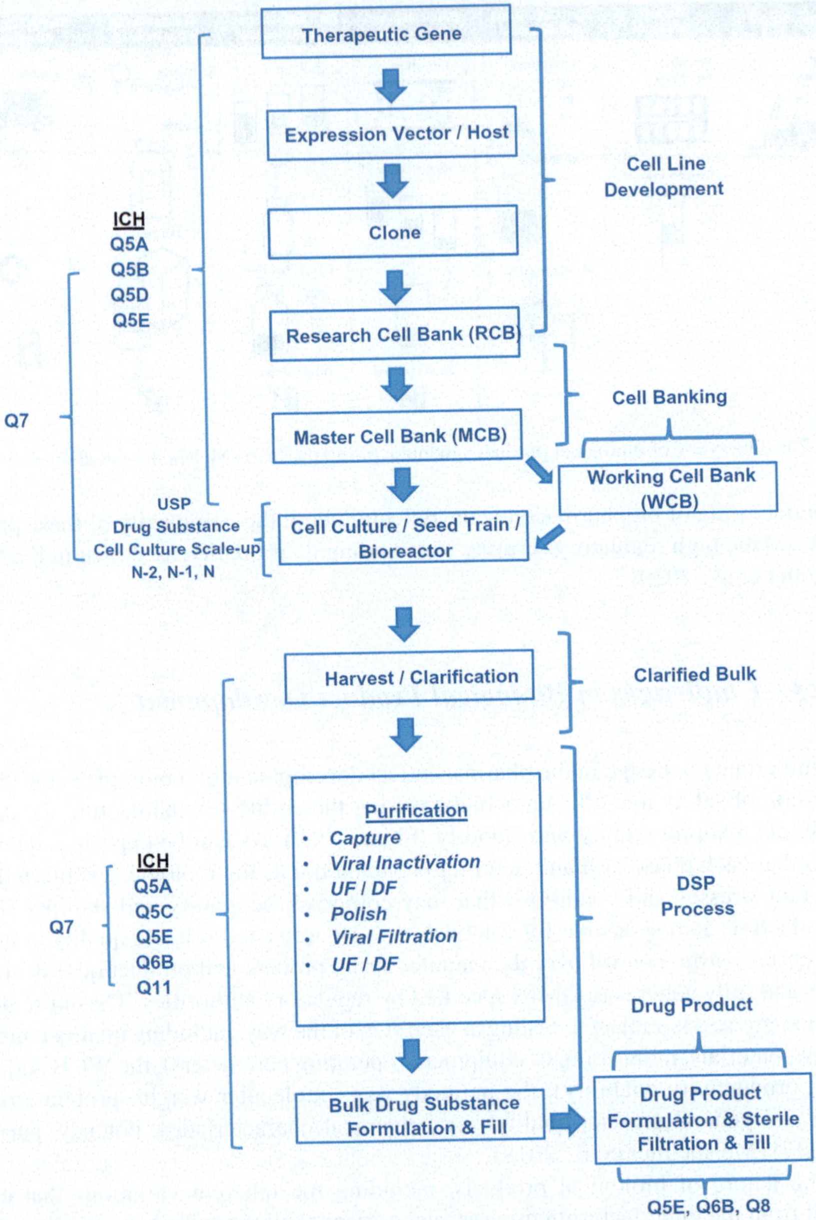

Fig. 7.7 Typical biologic manufacturing process flow diagram with appropriate ICH guidance per steps. (Conner et al., 2020) (reprinted with permission from Elsevier)

However, changes in the manufacturing process, equipment, or facilities could result in changes in the biological product itself and sometimes require additional clinical studies to demonstrate the product's safety, identity, purity, and potency.

Biologics are often extremely sensitive to physical conditions and enzymatic action and require complex bioassays. What can be done to lower their costs of goods?

7.3 Enhancing Value Proposition in Biologics Development

As discussed in Chap. 5, from a business strategy perspective, the pharmaceutical industry has a profitable but also extremely risky business model. Also, as shown in Fig. 5.2, until the product launch, the cash flow is negative due to the R&D expenses. Upon launch, the drug sales bring in revenues, and cash flow becomes positive. However, the cash flow is severely impacted as soon as the drug's exclusivity is lost, allowing generics to enter the market. In the case of biologics product development, the impact of various stages, such as Incubation, Growth, Maturity, and Decline, on the value proposition of the biologics need to be better understood, and the opportunities to enhance the product lifecycle needs to be evaluated.

In 2015, analysts from Deloitte released a report that evaluated the current global biopharmaceutical portfolio. This report ponders on the increased therapeutic competition, a greater prevalence of large molecule drugs, expansion in the number of personalized or targeted products, and a rise of treatments for many orphan diseases. These trends have given rise to biopharmaceutical products with extremely limited production runs, highly specific manufacturing requirements, and genotype-specific products. This fundamental shift in the overall product mix and a focus on continuing to improve the efficiency and effectiveness of production is spurring an evolution in the technologies and processes needed to support advanced biopharmaceutical manufacturing (Jacoby et al., 2015). As per the report, biopharmaceutical manufacturers are generally making investments in the following areas:

- Continuous manufacturing to improve scalability and facilitate time to market while lowering capital and operating costs and enhancing quality
- New process analytical tools to improve process robustness, accelerate scale-up to commercial production, and drive more efficient use of resources
- Single-use systems to increase flexibility and reduce production lead times while lowering capital investment and energy requirements
- Alternative downstream processing techniques to improve yields while lowering costs, green chemistry to reduce waste, and new vaccine and therapy production methods to increase capacity, scalability, and flexibility.
- It was increasing the effectiveness of medicines and supporting patient compliance by developing products that reflect improvements in drug delivery systems and drug-device combination products.

The analysts concluded that innovation in manufacturing technology is helping to drive improved economics, flexibility, and quality while potentially benefiting patients both directly and indirectly. The rise of biopharmaceutical advanced manufacturing technologies is positively impacting society by benefiting patients, the environment, and the nation's standing as a leader in innovation and overall competitiveness (Jacoby et al., 2015). Therefore, it is necessary to understand and evaluate these new areas of development to fully appreciate their impact on the product development lifecycle.

7.3.1 *Improved Manufacturing Efficiency Through Continuous Manufacturing*

The commercial success of monoclonal antibodies led to significantly increased demand in their production scale. As discussed in Sect. 7.2, biopharmaceuticals require highly complex and costly manufacturing. A significant percentage of the high consumer cost of these products is due to their complex manufacturing. Manufacturers must maximize manufacturing efficiencies to minimize costs, decrease costs (optimize profits), and sell their products in an increasingly competitive world market. Therefore, in theory, bioprocessing efficiency can affect sales prices and, thus, access to these expensive products.

The manufacturing process of biopharmaceuticals, such as monoclonal antibodies, is divided into upstream (USP) and downstream processing (DSP), as shown in Fig. 7.8. The production of the monoclonal antibody in bioreactors using mammalian cells as an expression host and the separation of the liquid phase from the cells using centrifuges or filters is defined as USP. The subsequent DSP is designed to separate side components like host cell proteins (HCP) or host cell DNA (hDNA) from the main component. The most common unit operations used in the DSP are typically chromatography and filtration (Kornecki et al., 2017).

In biopharmaceutical manufacturing, titers, measured in mass (grams) of desired proteins per volume (liter) produced in the bioreactor,[2] and yields, measured as percent of mass (grams) of purified product obtained vs. mass (grams) at the start of purification, are the key benchmarks that manufacturers use to determine operational efficiency and improvements in bioprocessing. Titer is essentially the amount of protein produced in each liter of bioreactor fluid. If the titer doubles, then all else being equal, only half as much fluid volume needs to be purified, or half as many lots/batches are needed to produce the same amount of product. Thus, titer is a very important measure of the efficiency of a product's manufacturing and related costs. Similarly, yield can be considered how much (percent) of the protein coming out of the bioreactor is finally obtained after filtration and purification steps are completed. This is a measure of the efficiency a manufacturer has achieved in the downstream purification and filtration operations. In the 1980s and early 1990s, average titers at commercial scale started out at <0.5 g/L. The current average reported commercial-scale titer is 2.56 g/L. Researchers estimate that ≥3 g/L is now the industry standard titer for new bioprocesses being developed, with ≤7 g/L now presumed to be the

[2] Typical commercial production facilities have a scale of 10,000–25,000 (Jacoby et al., 2015).

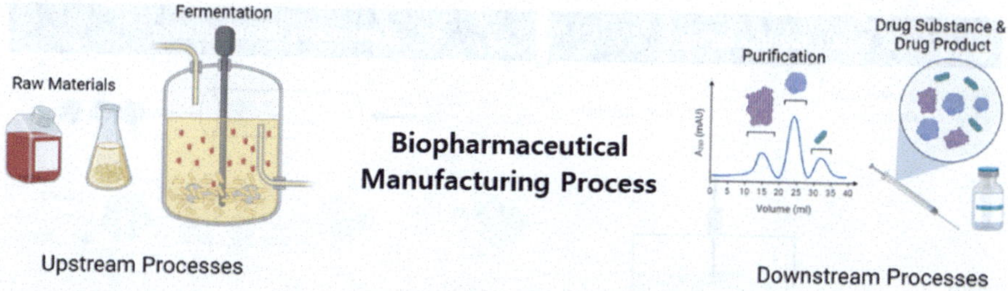

Fig. 7.8 Upstream and downstream biologic manufacturing process. (Adapted from Gerzon et al. (2022))

general industry top-end titer level that, while not unusual, is not often achieved. In terms of yields, researchers found a 70% yield to be the current industry average yield (Rader & Langer, 2014).

Most biopharmaceutical drugs are manufactured in batches in which human intervention is required to process a set quantity of material to be produced at the same time. Such operations were reasonable in the early phases of the industry but are inefficient and may be unsustainable as the global demand for these drugs grows. An alternative manufacturing approach that relies less on human labor and transitioning steps between unit operations requires a smaller facility footprint and is more amenable to scaling, automation, and adaptation across different drug modalities, which is, therefore, desirable. Continuous manufacturing is such an alternative that is gaining increasing popularity in the pharmaceutical industry (*CM was discussed in the context of small molecules-based product development in* Sect. 6.3.2).

Continuous operation, in the literal sense, would proceed without pauses. In a continuous manufacturing (CM) system, raw materials are continuously fed into a process train, while finished product materials are continuously removed at the other end (Fig. 7.9). In a more practical or transitional sense, a continuous process may include periodic or cyclic operations. CM requires a smaller footprint and relies less on labor than batch manufacturing. Many industries, such as the chemical, petrochemical, food, and mechanical industries, have transitioned from batch to continuous manufacturing to lower costs while addressing growing demands (Khanal & Lenhoff, 2021).

In the specific case of upstream processes, the enabling technology for continuous cell culture is a perfusion bioreactor that increases cell titers. Fresh media is fed, and depleted media is removed at the same rate. The removed media is then routed to a cell retention device to extract the harvested cell-culture fluid and recycle the cells back into the perfusion reactor. Additionally, a cell bleed is used to remove dead cells and improve cell viability. Given the continuous flow across a perfusion bioreactor, the cellular mean residence time is shorter and the distribution narrower, rendering the product quality more consistent than in batch and fed-batch bioreactors. Apart from greater consistency in cell viability, higher productivity is seen in perfusion culture than in fed-batch. There is also less buildup of undesirable metabolites and other impurities. Given the shorter product residence time inside a

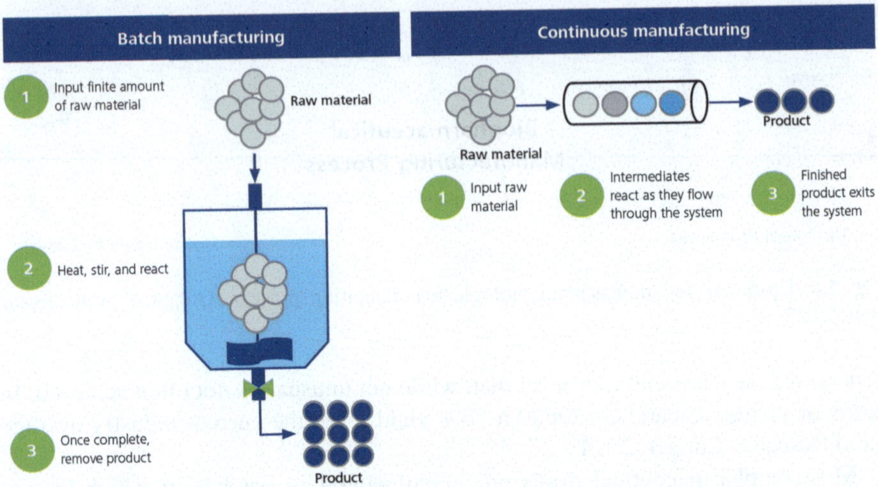

Fig. 7.9 Batch vs. continuous manufacturing process. (Jacoby et al., 2015)

perfusion bioreactor, it has become a preferred cell-culture method for producing degradation-prone therapeutics, and this continuous upstream operation has been incorporated into many commercial processes. Technological efforts in continuous unit operations, integrated processes, and process analytical technologies (PATs) that facilitate the implementation of continuous biomanufacturing have also been reported in the literature (Khanal & Lenhoff, 2021).

7.3.2 Better Process Control Through PAT

In 2004, the FDA issued a guidance entitled "PAT – A Framework for Innovative Pharmaceutical Development, Manufacturing, and Quality Assurance" that encouraged the industry to modernize manufacturing through enhancements in process control. The scientific, risk-based framework outlined in this guidance, Process Analytical Technology or PAT, is intended to support innovation and efficiency in pharmaceutical development, manufacturing, and quality assurance (FDA, 2004). The guidance provides a mechanism to design, analyze, and control pharmaceutical manufacturing process through the measurement of Critical Process Parameters (CPPs) with Critical Quality Attributes (CQAs) of the product. This encourages building quality into the design of the product and promoting the development of PAT solutions to digitize and provide real-time control of various critical parameters that influence product output and quality (Gerzon et al., 2022).

In an ideal manufacturing scenario, when a process is launched, its progress should be easily monitored and controlled until that process is completed. This means that at any given time, the condition and the quality of the product with respect to yield are transparent and known. This knowledge is lacking in current pharmaceutical manufacturing due to a lack of information feedback during the manufacturing process. This knowledge gap can be addressed by PAT solutions that report the condition of a product in real-time and facilitate the introduction of a feedback loop to control a process based on the reported parameters. Such an approach allows testing the quality of the product at different manufacturing stages to ensure quality by design (QbD) from raw materials to intermediate and to the final drug product. PAT has developed a broad suite of tools, including analytical, chemical, physical, microbiological, and risk analysis. Researchers have developed various methods of implementing PAT tools into bioprocess monitoring (Fig. 7.10) such as: (a) On-line analysis includes sample cycling from the process and measured using an analytical tool to provide real-time data, (b) In-line analysis implements the analytical tool in the bioprocess to provide data in real-time as the process is occurring, (c) Off- line analysis requires sampling from the bioprocess and transporting it for manual measurement in a different laboratory (i.e., Quality Control), and (d) At-line analysis requires manual sampling from the bioprocess and measuring the sample next to the bioprocess site, to improve acquisition time for the results.

The major goals of PAT integration into the manufacturing process can be summarized as the following three: (1) the reduction of product cycle times via real-time measurement and control systems, (2) the reduction of waste and reworks of the final product, and (3) the possibility for real-time product release. PAT tools for bioprocess monitoring include various analytical tools such as Near-Infrared Spectroscopy (NIR), Mid-Infrared Spectroscopy (MIR), Nuclear Magnetic Resonance (NMR), Raman Spectroscopy, and Differential Light Scattering (DLS) for monitoring different types of biological products throughout upstream and downstream production processes.

7.3.3 *Potential for Digital Manufacturing*

From the early-age machining process to today's automated manufacturing industries, manufacturing has achieved huge advancements. Going through various inventions and repetitive trials over the years (starting with the Industrial Revolution of the late 1700s), manufacturing and industrialization systems have now reached their fourth generation of the Industrial Revolution (Fig. 7.11). In 2011, a group of business leaders and academic experts introduced the term Industry 4.0 as the fourth industrial revolution to enhance the manufacturing industry. This concept of Industry 4.0 builds heavily on interconnectivity through artificial intelligence, machine learning, and the processing of real-time data (Phuyal et al., 2020). Like the revolutions that preceded it, Industry 4.0 has the potential to raise global income levels and improve the quality of life for populations around the world.

Within Industry 4.0, smart manufacturing is a manufacturing approach that incorporates advanced technologies such as artificial intelligence, the Internet of Things (IoT), robotics, and big data analytics to optimize production processes and increase efficiency. The implementation of smart manufacturing has also given rise

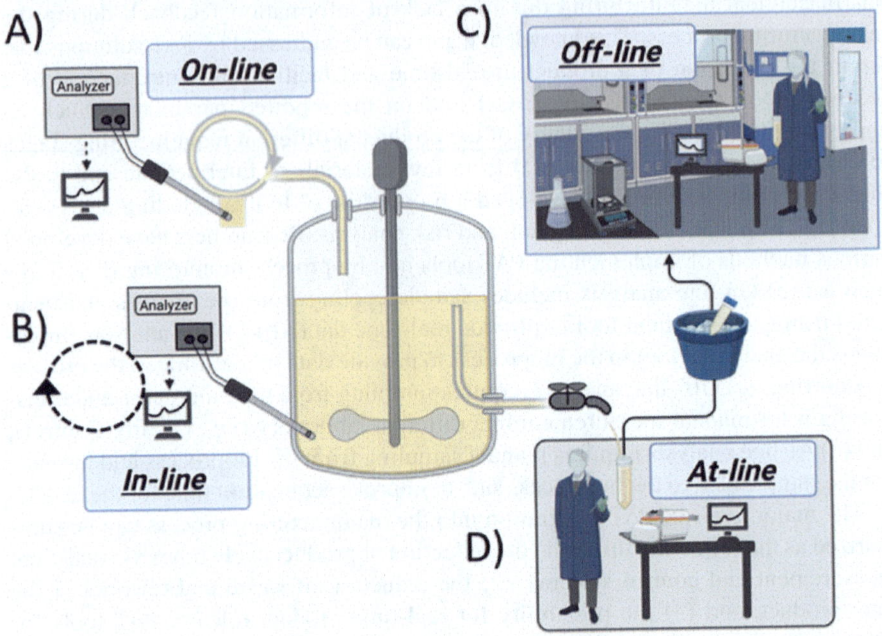

Fig. 7.10 Various methods of implementing PAT tools into bioprocess monitoring (Gerzon et al., 2022)

to the development of the concept of a Digital Twin. A digital twin is a virtual representation of a physical system or process that allows for real-time monitoring, analysis, and optimization. In the context of smart manufacturing, a digital twin can be used to simulate and optimize the production process, predict and prevent equipment failures, and improve the efficiency and quality of part production. The digital twin can provide a detailed, accurate representation of the physical object or system, including its behavior, performance, and interactions with the environment. Digital twins use machine learning, data analytics, and multi-physics simulation in order to simulate and analyze different working conditions and other factors that affect a system (Soori et al., 2023).

In the case of the biopharmaceutical manufacturing process, the digital twin approach can be developed using PAT. Implementation of PAT tools helps in creating processes with feedback loop control, which allows for better process understanding, control, and optimization. By understanding CPPs that influence bioprocesses, that knowledge can be adapted and optimized for use in various platforms. Thus, when generating large amounts of data, a digital twin could be created to provide information about a process and expected data readout with associated software, simulate a manufacturing process, and change and adapt equipment and procedures for a new platform. Thus, the digital copy containing all the critical information about a process can be transferred from various plants or physical

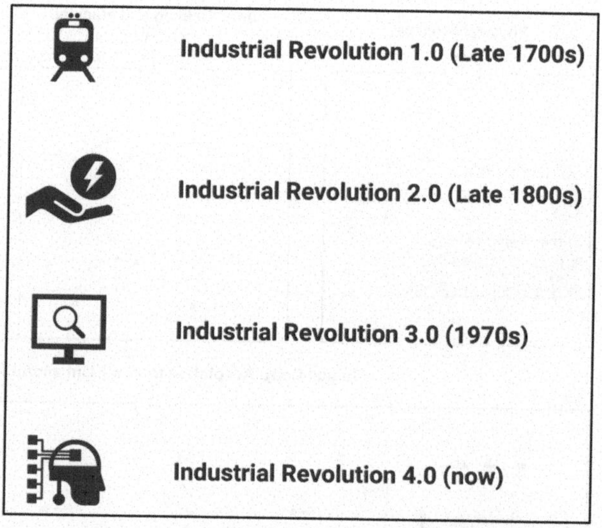

Fig. 7.11 Various industrial revolutions in human history

locations to a new process. This concept is demonstrated in Fig. 7.12. To date, there are not any digital twins being directly applied in biopharmaceutical applications. However, researchers are developing models to create digital twin frameworks that can be used for bioprocesses in biopharmaceutical manufacturing (Gerzon et al., 2022).

7.3.4 Improving Productivity Through Single-Use Systems

The current standard manufacturing scheme for biologics is optimized for fed-batch bioreactors that are well adapted for suspension cell culture processes, followed by a combination of different filtration and chromatography unit operations to achieve the target purity requirement and yield. Both fed-batch bioreactors and resin column chromatography operations experience an oversizing problem when scaled up to pilot and manufacturing process scales. Fed-batch reactors, which comprise about 90% of commercial biologics production, require large-volume tanks to counter the low productivity per unit volume, and resin chromatography columns, limited by intrinsically slow mass transfer rates, have to be over-sized in order to operate at a higher flow rate to achieve desirable productivity. This approach of attaining better productivity via enlarged working units occurs at the expense of process economics and flexibility. Large reactors (up to 25,000 L) and chromatography columns (up to 2 m in diameter with 10 cm to 20 cm bed height) directly increase the facility cost due to increased building and equipment expenses as well as associated piping and hardware costs (including preparing, holding, and cleaning). In addition to the economic disadvantage, over-sized bioprocess plants offer limited flexibility, which directly contradicts the emerging trend in the biopharmaceutical industry of

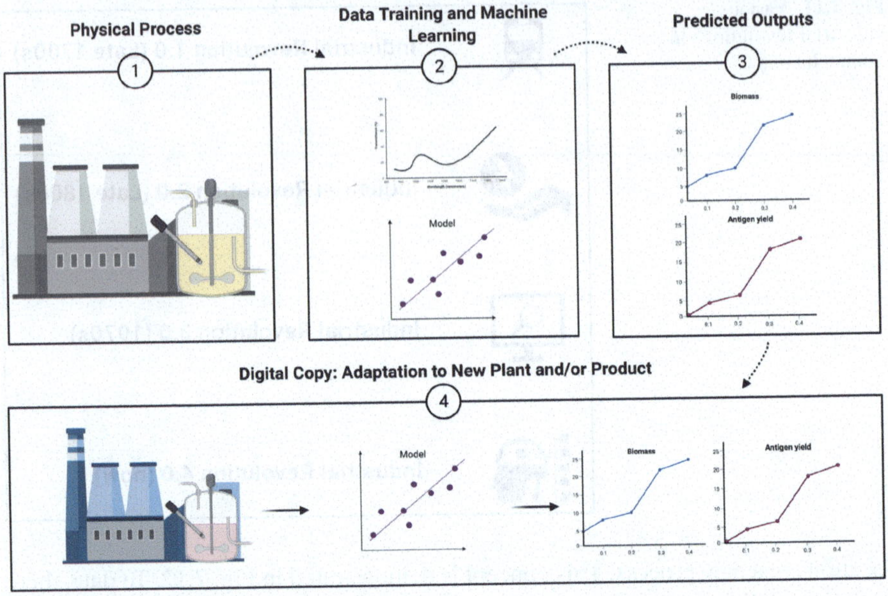

Fig. 7.12 Schematic design for the implementation of a digital twin in the biopharmaceutical industry. (Gerzon et al., 2022)

production on demand. Moreover, the industry is transitioning from large, single-product facilities to scalable, multi-product facilities for greater product variety. The emergence of personalized medicine, as well as smaller product campaigns associated with orphan drugs, biomarkers, and smaller disease paradigms, puts more pressure on manufacturing volume management. All these scenarios, coupled with the unpredictability of demand, necessitate facility flexibility (Jacquemart et al., 2016).

In recent years, the use of single-use systems, such as single-use culture bags in place of conventional stainless steel culture tanks, has been increasing in the manufacture of biologics. Single-use stirred tank bioreactors, such as GE Healthcare's Xcellerex bioreactors that were introduced in the mid-to-late 2000s, are now commonly operated in small- to mid-scale bioprocessing projects. In this reactor, cells are cultured in a replaceable plastic bag that is housed within a stainless-steel tank. Mechanical agitation is supported through the bottom, a center-mounted magnetic drive that couples with the impeller in the bag. Due to weight limitations, single-use stirred tank bioreactors above 2000 L are not feasible. However, over the last decade, there has been a dramatic increase in expression yield from progress in cell lines, expression systems, and culture. These advances have made large-scale biomanufacturing with the Xcellerex bioreactors more practical. For instance, 100 kg/year of mAb can be manufactured in a few single-use 2000 L bioreactors operated continuously or in parallel, whereas multiple ≥10,000 L bioreactors were required for the same output only a decade ago (Jacquemart et al., 2016).

Apart from the increased efficiency of manufacturing, the utilization of novel single-use technologies has other promising advantages that include the prevention of cross-contamination in the manufacture of high-potency compounds, flexibility to manufacture multiple products, and elimination of the need for cleaning and steam sterilization, including those during validations. However, this approach has also raised new issues and concerns related to process control and product quality assurance. In order to ensure the safety of biologics when employing such single-use systems in the manufacturing process, various issues need to be considered, including possible extractables/leachables and particles arising from the components used in single-use systems. (Ishii-Watabe et al., 2015). One other challenge of using single-use systems is the disposal of plastics. Due to the nature of the products that are produced, standard recycling practices are not yet prevalent in the industry. As a result, some of the environmental benefits of single-use systems are countered by the need to use incineration or landfills to dispose of contaminated plastic. However, many suppliers are currently developing recycling programs to address this issue (Jacoby et al., 2015).

7.3.5 *Understanding the Challenges from Biosimilars*

As discussed in the previous section, an important and persistent aspect of biologics drug products is their high prices that stem from the complex development and manufacturing process. Another reason for the high prices of biologics is that they are novel, first-in-class entities that enjoy virtual monopolies with the ability to command high prices. Biologics can obtain patent protection, which lasts for 20 years from the date the patent application is filed. Questions exist about the degree of protection afforded by biological patents. The Biologics Price Competition and Innovation Act (BPCIA) of 2009 recognizes the importance of encouraging innovation, but it also provides a pathway for competition once monopoly protection ends[3] (Carl et al., 2022). Accordingly, the BPCIA provides a 12-year market exclusivity and a 4-year data exclusivity beginning when the biologic drug receives FDA marketing approval. Each exclusivity can be extended to 6 months for pediatric applications, as seen in Table 5.5. A biosimilar cannot be marketed until the 12-year exclusivity expires. These exclusivity protections are intended to encourage biological R&D. Patents can be challenged in court, but exclusivity cannot. All these factors limit the number of biosimilars in the marketplace.

A biosimilar is a biological product that is highly similar to and has no clinically meaningful differences from an existing FDA-approved reference product.[4] There are two types of biosimilars: (1) biosimilar products and (2) interchangeable

[3] In the European Union, a similar approval pathway was introduced in 2005 (Carl et al., 2022).

[4] A reference product is the single biological product, already approved by FDA, against which a proposed biosimilar product is compared. A reference product is approved based on, among other things, a full complement of safety and effectiveness data.

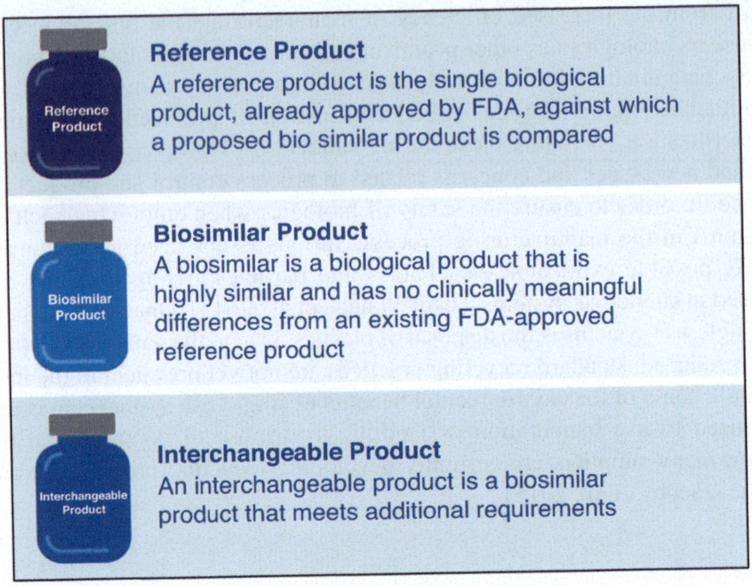

Fig. 7.13 Definitions of various biological products. (FDA, 2023a)

products. The difference between the various types of biological products is illustrated in Fig. 7.13.

There are two criteria for biosimilars to be approved by the FDA. These two criteria require the prospective biosimilar to be (1) highly similar and (2) no clinically meaningful differences from the reference product. In order to demonstrate that a potential biosimilar is highly similar to a reference product, the manufacturer developing the biosimilar extensively characterizes the structure and function of both the reference product and the proposed biosimilar. Minor differences between the reference product and the proposed biosimilar product in clinically inactive components are acceptable. For example, these could include minor differences in the stabilizer or buffer compared to what is used in the reference product. Any differences between the proposed biosimilar product and the reference product are carefully evaluated during the regulatory review process (FDA, 2023a). A manufacturer must also demonstrate that its proposed biosimilar product has no clinically meaningful differences from the reference product in terms of safety, purity, and potency (safety and effectiveness). This is generally demonstrated through human pharmacokinetic and pharmacodynamic studies, an assessment of clinical immunogenicity, and, if needed, additional clinical studies (FDA, 2023a).

An interchangeable product is a biosimilar product that meets additional requirements outlined by the BPCIA.[5] As part of fulfilling these additional requirements,

[5] The BPCIA of 2009 amends the PHSA to create an abbreviated approval pathway, called 351(k) for biological products shown to be biosimilar to, or interchangeable with, an FDA-licensed reference biological product. The BPCIA is closely related to the Drug Price Competition and Patent

information is needed to show that an interchangeable product is expected to produce the same clinical result as the reference product in any given patient. Also, for products administered to a patient more than once, the risk in terms of safety and reduced efficacy of switching back and forth between an interchangeable product and a reference product will have been evaluated. An interchangeable product may be substituted for the reference product without the involvement of a healthcare provider who prescribed the product (FDA, 2023a).

The development of biosimilars requires significant resource investment in establishing manufacturing processes and analytical method development to enable production and comparative characterization of the molecular structure and quality attributes of the target reference product and its copy. The early phase of biosimilar development is more extensive than the early development of novel biologics because it involves extensive analytical characterization of two products—the innovator and the biosimilar candidate. It may also be quite costly because of the necessity of purchasing a large quantity of the reference product, often manufactured at different sites and marketed as multiple versions in different countries, to be used in comparative evaluation. Thus, CMC work on biosimilars requires not only the development of a consistent and well-controlled manufacturing process of the biosimilar product itself but also extensive comparative side-by-side analyses of procured reference product lots and manufactured biosimilar candidate batches. Once the analytical characterization of the intended biosimilar candidate and the reference product has demonstrated a "highly similar quality profile", the extent of non-clinical and clinical studies with the biosimilar candidate should be reduced compared to the novel biopharmaceuticals as the clinical efficacy and safety of the reference product are well understood (Herzyk, 2013).

On March 6, 2015, the FDA approved filgrastim-sndz (Zarxio®; Sandoz/Novartis), the first biosimilar ever to receive approval in the United States. Filgrastim-sndz is biosimilar to filgrastim (Neupogen®; Amgen), which was originally approved by the FDA in 1991. The formulation of filgrastim-sndz differs from that of filgrastim by one inactive component. Filgrastim-sndz is approved for the same five indications as its reference drug filgrastim (Raedler, 2016). Since 2015, there has been a significant increase in the number of biosimilars approved by the FDA. According to the Purple Book,[6] as of November 2023, there are currently 44 approved biosimilars in the US. The breakdown of these biosimilars by their reference product is shown in Fig. 7.14. The number of biosimilars approved each year is steadily increasing and averages about five approvals per calendar year (Fig. 7.15).

Term Restoration Act of 1984 (or referred to as the "Hatch-Waxman Act"), which established abbreviated pathways for the approval of drug products under the FDCA.

[6]The Purple Book database contains information on all FDA-licensed (approved) biological products regulated by the Center for Drug Evaluation and Research (CDER), including licensed biosimilar and interchangeable products, and their reference products. The Purple Book also contains information about all FDA-licensed allergenic, cellular and gene therapy, hematologic, and vaccine products regulated by the Center for Biologics Evaluation and Research (CBER).

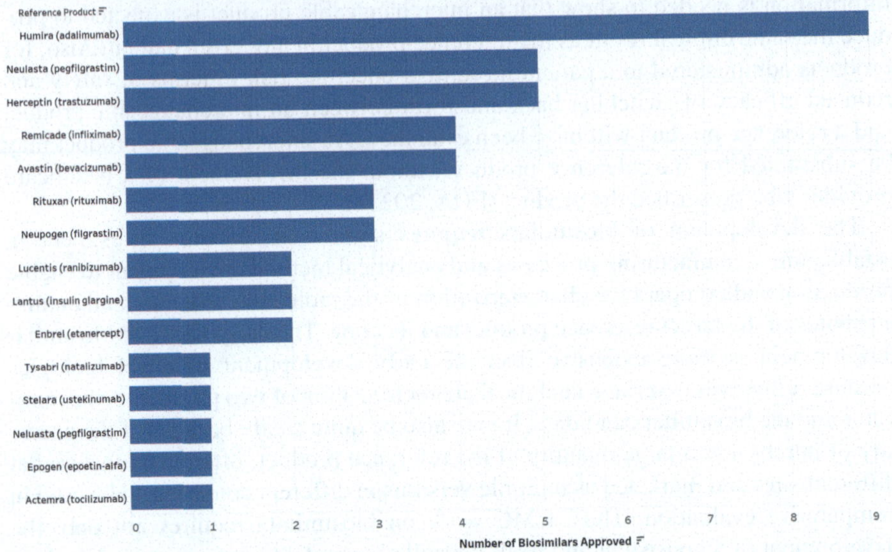

Fig. 7.14 Number of biosimilars approved in the USA (Data Source: FDA's purple book (FDA, 2023b))

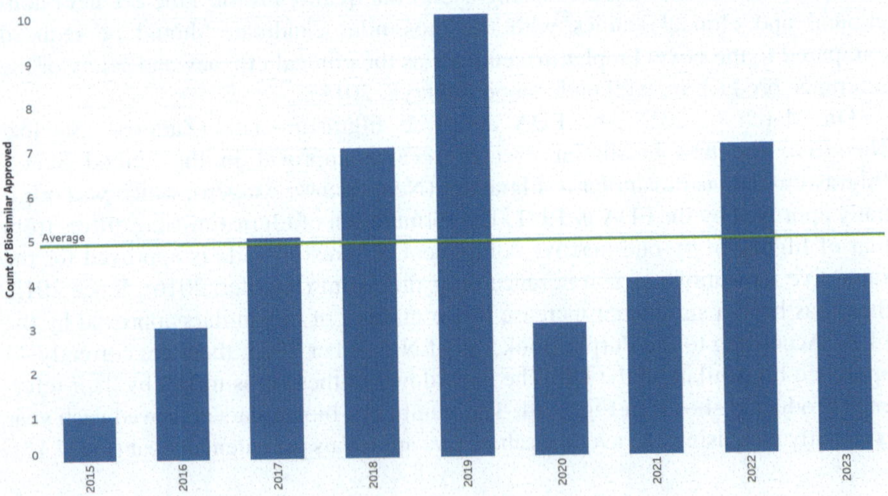

Fig. 7.15 Number of biosimilars approved in the USA per year (Data Source: FDA's purple book (FDA, 2023b))

The impact of biosimilars in the marketplace is substantial. In a recent study, researchers assessed the uptake and prices of biosimilars in the United States compared with 2 European countries (Germany and Switzerland) with national mechanisms for drug price negotiation. The study cohort included 15 biosimilars and six

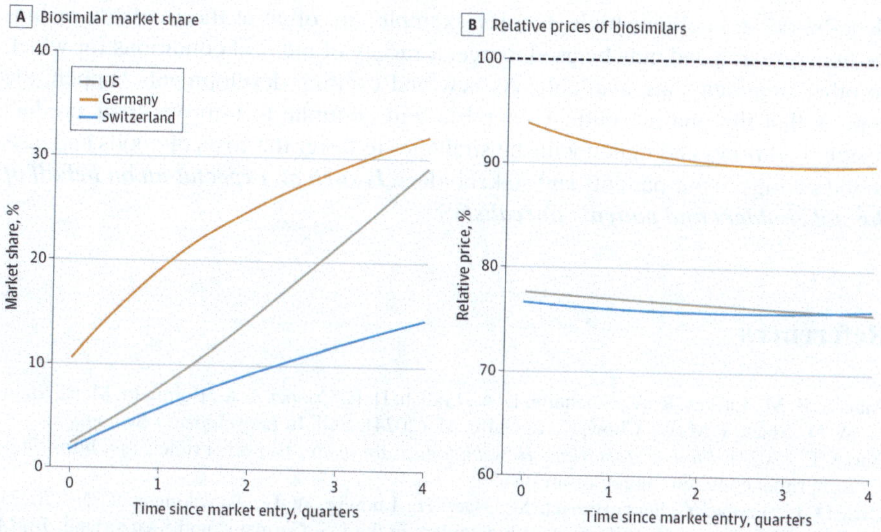

Fig. 7.16 Relative prices of biosimilars vs reference products (Biologics) in the USA, Germany, and Switzerland. (Carl et al., 2022)

biologics for the US, 52 biosimilars and 15 biologics for Germany, and 28 biosimilars and 13 biologics for Switzerland. The researchers concluded that the uptake of biosimilars increased over time in all countries. On average, the biosimilar market share at launch was highest in Germany; however, it increased at the fastest rate in the US. In the US, the relative prices of biosimilars were lower at market entry compared with their reference products, with a variation between 55% and 90% and a slight decrease over time. In Germany, most biosimilars entered the market with relative prices between 65% and 100%. In Switzerland, prices for biosimilars were lower at market entry compared with their reference products, with relative prices ranging between 70% and 80%, as shown in Fig. 7.16 (Carl et al., 2022).

7.4 Summary—What Else Is Possible—Part 2?

The future of biologics is bright. As per industry thought leaders, while global regulatory agencies have focused much of their attention on approving COVID-19 tests and vaccines for use over the last few years, now the focus is shifting to other regular novel drug approval processes. The significant number of new approvals raises questions about additional global capacity and whether capacity constraints will be felt if they are not addressed in a timely manner by the industry. To bring biotherapeutics to realization, a significant amount of planning, capital expenditure, access to talented R&D and CMC personnel, and extensive experience in biologics production are required. The emergence of novel modalities also presents challenges in terms of technology development and supply-chain issues that will require companies to make substantial investments in manufacturing capacity (Dakhole, 2023).

Gene-based and cellular biologics, for example, are often at the forefront of biomedical research and may be used to treat a variety of medical conditions for which no other treatments are available. As new and exciting developments happen, the hope is that the pharmaceutical scientists will continue to improve their product design workflows and manufacturing strategies to lower the costs of goods and pass on the savings to our patients and stakeholders. ***Is such an expectation on behalf of the stakeholders and patients unrealistic?***

References

Alberts, B. M., Laskey, R. A., Staehelin, L. A., Lodish, H. F., Cooper, J. A., Bernfield, M. R., Stein, W. D., Slack, J. M. W., Chow, C., & Cuffe, M. (2024). Cell. In *Encyclopedia Britannica*.

Burke, E. (2023). *Biomanufacturing: How biologics are made*. Biotech Primer, Inc. https://biotechprimer.com/how-biologics-are-made/

Carl, D. L., Laube, Y., Serra-Burriel, M., Naci, H., Ludwig, W.-D., & Vokinger, K. N. (2022). Comparison of uptake and prices of biosimilars in the US, Germany, and Switzerland. *JAMA Network Open, 5*(12), e2244670. https://doi.org/10.1001/jamanetworkopen.2022.44670

Conner, J., Wuchterl, D., Lopez, M., Minshall, B., Prusti, R., Boclair, D., Peterson, J., & Allen, C. (2020). Chapter 31—The biomanufacturing of biotechnology products. In C. Shimasaki (Ed.), *Biotechnology entrepreneurship* (2nd ed., pp. 447–480). Academic. https://doi.org/10.1016/B978-0-12-815585-1.00031-0

Dakhole, V. (2023). *The rise of biologics: is the sector ready for future demand?* Retrieved March 6, 2024, from https://pmlive.com/intelligence/the_rise_of_biologics_is_the_sector_ready_for_future_demand_1495996/#:~:text=Future%20outlook,substantial%20investments%20in%20manufacturing%20capacity

de la Torre, B. G., & Albericio, F. (2022). The pharmaceutical industry in 2021. An analysis of FDA drug approvals from the perspective of molecules. *Molecules, 27*(3). https://doi.org/10.3390/molecules27031075

Edgman-Levitan, S., & Schoenbaum, S. C. (2021). Patient-centered care: Achieving higher quality by designing care through the patient's eyes. *Israel Journal Health Policy Research, 10*(1), 21. https://doi.org/10.1186/s13584-021-00459-9

Etheredge, L. (2024). Henrietta Lacks. In *Encyclopedia Britannica*.

FDA. (2004). *PAT—A framework for innovative pharmaceutical development, manufacturing, and quality assurance: Guidance for industry*. Retrieved from https://www.fda.gov/regulatory-information/search-fda-guidance-documents/pat-framework-innovative-pharmaceutical-development-manufacturing-and-quality-assurance

FDA. (2018a). *Patient-focused drug development glossary*. FDA. Retrieved February 28, 2024 from https://www.fda.gov/drugs/development-approval-process-drugs/patient-focused-drug-development-glossary

FDA. (2018b). *What are "biologics" questions and answers*. Retrieved November 6 from https://www.fda.gov/about-fda/center-biologics-evaluation-and-research-cber/what-are-biologics-questions-and-answers

FDA. (2023a). *Biological product definitions*. Retrieved from https://www.fda.gov/files/drugs/published/Biological-Product-Definitions.pdf

FDA. (2023b). *Biosimilar product information*. https://www.fda.gov/drugs/biosimilars/biosimilar-product-information

Gerzon, G., Sheng, Y., & Kirkitadze, M. (2022). Process analytical technologies—Advances in bioprocess integration and future perspectives. *Journal of Pharmaceutical and Biomedical Analysis, 207*, 114379. https://doi.org/10.1016/j.jpba.2021.114379

Greenwood, J. C. (2014). Chapter 1—Unleashing the promise of biotechnology to help heal, fuel, and feed the world. In C. Shimasaki (Ed.), *Biotechnology entrepreneurship* (pp. 3–13). Academic. https://doi.org/10.1016/B978-0-12-404730-3.00001-4

Herzyk, D. J. (2013). Chapter 5—Overview of biosimilar therapeutics. In L. M. Plitnick & D. J. Herzyk (Eds.), *Nonclinical development of novel biologics, biosimilars, vaccines and specialty biologics* (pp. 141–157). Academic. https://doi.org/10.1016/B978-0-12-394810-6.00005-8

Institute_of_Medicine. (2001). *In crossing the quality chasm: A new health system for the 21st century*. National Academies Press (US) Copyright 2001 by the National Academy of Sciences. All rights reserved. https://doi.org/10.17226/10027.

Ishii-Watabe, A., Hirose, A., Katori, N., Hashii, N., Arai, S., Awatsu, H., Eiza, A., Hara, Y., Hattori, H., Inoue, T., Isono, T., Iwakura, M., Kajihara, D., Kasahara, N., Matsuda, H., Murakami, S., Nakagawa, T., Okumura, T., Omasa, T., et al. (2015). Approaches to quality risk management when using single-use systems in the manufacture of biologics. *AAPS PharmSciTech, 16*(5), 993–1001. https://doi.org/10.1208/s12249-015-0368-z

Jacoby, R., Pernenkil, L., Harutunian, S., Heim, M., & Sabad, A. (2015). *Advanced biopharmaceutical manufacturing: An evolution underway*. https://www2.deloitte.com/content/dam/Deloitte/us/Documents/life-sciences-health-care/us-lshc-advanced-biopharmaceutical-manufacturing-white-paper-051515.pdf

Jacquemart, R., Vandersluis, M., Zhao, M., Sukhija, K., Sidhu, N., & Stout, J. (2016). A single-use strategy to enable manufacturing of affordable biologics. *Computational and Structural Biotechnology Journal, 14*, 309–318. https://doi.org/10.1016/j.csbj.2016.06.007

Kadam, R. A. (2017). Informed consent process: A step further towards making it meaningful! *Perspectives in Clinical Research, 8*(3), 107–112. https://doi.org/10.4103/picr.PICR_147_16

Khanal, O., & Lenhoff, A. M. (2021). Developments and opportunities in continuous biopharmaceutical manufacturing. *MAbs, 13*(1), 1903664. https://doi.org/10.1080/19420862.2021.1903664

Kornecki, M., Mestmäcker, F., Zobel-Roos, S., Heikaus de Figueiredo, L., Schlüter, H., & Strube, J. (2017). Host cell proteins in biologics manufacturing: The good, the bad, and the ugly. *Antibodies, 6*(3), 13. https://www.mdpi.com/2073-4468/6/3/13

Li, Y., Tang, W., & Guo, M. (2021). The cell as matter: Connecting molecular biology to cellular functions. *Matter, 4*(6), 1863–1891. https://doi.org/10.1016/j.matt.2021.03.013

Liu, J. K. (2014). The history of monoclonal antibody development – Progress, remaining challenges and future innovations. *Ann Med Surg (Lond), 3*(4), 113–116. https://doi.org/10.1016/j.amsu.2014.09.001

Makurvet, F. D. (2021). Biologics vs. small molecules: Drug costs and patient access. *Medicine in Drug Discovery, 9*, 100075. https://doi.org/10.1016/j.medidd.2020.100075

Melsheimer, R., Calmann, M., DeRitis, A., Philip, V., Van Gog, F., Doolittle, L., Goyal, K., & Neblock, D. (2018). Ensuring product quality, consistency and patient supply over time for a large-volume biologic: Experience with Remicade®. *BioDrugs, 32*(5), 405–414. https://doi.org/10.1007/s40259-018-0300-7

Morrow, T., & Felcone, L. H. (2004). Defining the difference: What makes biologics unique. *Biotechnology Healthcare, 1*(4), 24–29.

Phuyal, S., Bista, D., & Bista, R. (2020). Challenges, opportunities and future directions of smart manufacturing: A state of art review. *Sustainable Futures, 2*, 100023. https://doi.org/10.1016/j.sftr.2020.100023

Rader, R. A., & Langer, E. S. (2014). Biopharmaceutical manufacturing: Historical and future trends in titers, yields, and efficiency in commercial-scale bioprocessing. *Bioprocessing Journal, 13*(4).

Raedler, L. A. (2016). Zarxio (Filgrastim-sndz): First biosimilar approved in the United States. *Am Health Drug Benefits, 9*(Spec Feature), 150–154.

Rao, S. K. (2011). Pricing biologics: Issues, strategic priorities and a conceptual model. *Journal of Commercial Biotechnology, 17*(1), 7–23. https://doi.org/10.1057/jcb.2010.30

Roy, A. (2019). *Biologic medicines: The biggest driver of rising drug prices*. Forbes. Retrieved November 23, 2023 from https://www.forbes.com/sites/theapothecary/2019/03/08/biologic-medicines-the-biggest-driver-of-rising-drug-prices/?sh=2d4ad0918b00

Sheets, R. (2018). Chapter 1—Introduction to the regulatory process for biologicals. In R. Sheets (Ed.), *Fundamentals of biologicals regulation* (pp. 5–18). Academic. https://doi.org/10.1016/B978-0-12-809290-3.00001-8

Shire, S. J. (2009). Formulation and manufacturability of biologics. *Current Opinion in Biotechnology, 20*(6), 708–714. https://doi.org/10.1016/j.copbio.2009.10.006

Skloot, R. (2010). *The immortal life of Henrietta lacks*. Broadway Books.

Soori, M., Arezoo, B., & Dastres, R. (2023). Digital twin for smart manufacturing: A review. *Sustainable Manufacturing and Service Economics, 2*, 100017. https://doi.org/10.1016/j.smse.2023.100017

Tariman, J. D. (2018). Biosimilars: Exploring the history, science, and progress. *Clinical Journal of Oncology Nursing, 22*(5), 5–12. https://doi.org/10.1188/18.Cjon.S1.5-12

Various. (2024). *Henrietta lacks*. Wikipedia. Retrieved February 26, 2024 from https://en.wikipedia.org/wiki/Henrietta_Lacks

Chapter 8
Sustaining Innovation in an Age of Disruption

Talent is Universal; Opportunity is Not

Rye Barcott

Contents

Abstract The high costs of prescription medicines are a major societal and political issue. In response to this social crisis, the legal and regulatory framework for pharmaceuticals is changing fast. Also, in this age of disruptive innovation, no idea for improving the innovation potential of our products should be off the table. Therefore, it is high time that pharmaceutical scientists who design, develop, and manufacture drug products are part of the solution to medicine accessibility. For example, if the awareness of the costs paid by the patients was factored into product development and was indeed a design parameter that would be scrutinized similar to how other design parameters (such as assay, impurities, shelf life, etc.) are integrated into the TPP and QTPP, then the overall attempts to bring balance to the global health inequity will be lot more wholesome and collectively owned by the designers of the product. Such an inclusive approach may also help improve patient-focused drug development efforts.

B. Mittal, *Sustainable Global Health Systems and Pharmaceutical Development*, AAPS Advances in the Pharmaceutical Sciences Series 60,
https://doi.org/10.1007/978-3-031-50415-0_8

Keywords Innovation matrix · Sustainable value creation · Juran trilogy · Cost of poor quality · QbED

8.1 Introduction

Pharmaceutical product development is a critical social requirement. However, with the fast pace of our ever-changing world, it is equally important for pharmaceutical product development to adapt. With the advent of disruptive technologies and evolving business models, the need to adapt is no longer wishful thinking but an important hallmark to sustain the innovation potential. In his influential book, *Mapping Innovation: A Playbook for Navigating a Disruptive Age*, Greg Satell provides a template for mapping the innovation space by proposing an aptly named tool that he calls the Innovation Matrix (Fig. 8.1).

The innovation matrix starts by asking two fundamental questions: (1) How well is the problem defined?, and (2) Who is best placed to solve it? The basic premise behind question #1 is that a well-intentioned and clear framing of any problem starts to allow for its associated innovation strategy to adapt. Take, for example, the healthcare debate, and the problem is "We want to reduce global health inequity". The framing of this question emphasizes the desire that we all as members of society want to see global health equality be accomplished, and we all are therefore

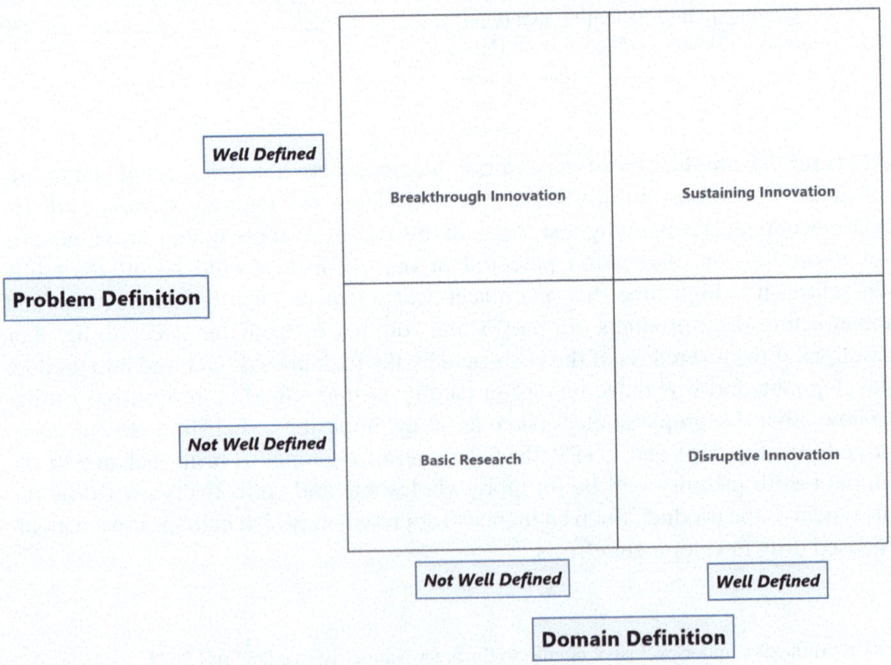

Fig. 8.1 The innovation matrix. (Adapted from Satell, 2017)

taking responsibility to see it through as opposed to passing the responsibility on to the government or a particular sector of our society. In a similar vein, question #2 posits that solutions exist to problems, but the intent to find these solutions has to be undertaken by people who are best positioned to implement these solutions. This approach allows for out-of-the-box thinking to develop and for critically evaluating the skill sets needed to solve problems. After all, who else but we, pharmaceutical scientists, are best positioned to solve the access to medication issues by building cost-effectiveness into our products' design and manufacturing? Satell claims that once any individual or organization starts asking these questions, they will find that the issues get clarified very quickly. There may not be a simple answer, but the exercise of asking these two questions as part of the innovation matrix is very helpful in defining an approach (Satell, 2017).

The four quadrants that make up the innovation matrix allow for narrowing down options for an innovation strategy. In the Basic Research quadrant, when the aim is to discover something completely new, neither the problem nor the domain is well defined. For the most part, basic research is funded by the government and undertaken by academic institutions. Fundamental research is likely to result in a broad range of applications. In the case of Breakthrough Innovation, the problems are well defined, although the expertise domain may not be well defined. Most organizations focus on solving problems that they can clearly define, although even well-defined problems can be hard to solve. Sustaining Innovation quadrant is based on the promise that every technology needs to get better. In this quadrant, the problem and the domain are well defined, and it is by integrating the different ideas and the desire to make a product better that the sustainability of the innovation potential can be maintained. In the last quadrant of Disruptive Innovation, after a certain point of sustained innovation, products and services may be created that is actually better than what the market is demanding, which opens up an opportunity for a competitor to change the basis of competition. Another way to view disruptive innovation is as an existing technology put toward a new purpose (Satell, 2017).

Once one starts to evaluate the Innovation Matrix critically, it becomes clear that this is a great tool for evaluating the approach that an organization may take in tackling the issues it is facing. Such approaches in the past have yielded amazing results for society where a combination of basic research, followed by breakthrough innovation and disruptive innovation, led to the ideal state of sustaining innovation. One such example is that of the Green Revolution of the 1960s.

8.1.1 Philosophical Digression—Bringing Prosperity by Tackling Hunger

The post-World War II years were a period of great economic prosperity in the United States. Unfortunately, the same was not true for the rest of the world. Most of Europe and Japan were destroyed due to the collateral damages of the War, but the rest of the world was not faring much better either. The immediate decade following World War II saw the horrifying impacts of a Nuclear Age that was shortly followed by the Cold War that ultimately culminated in further suffering through the Korean War, the Arab-Israeli conflicts, and the Vietnam War. Nevertheless, with all the chaos in the world, this was also a period of new thinking and tremendous innovation. One of which was the Green Revolution.

The Green Revolution, or the Third Agricultural Revolution, was a period of technology transfer initiatives that saw greatly increased crop yields (especially wheat and rice). These changes in agriculture began in developed countries in the early twentieth century and spread globally till the late 1980s. In the late 1960s, farmers began incorporating new technologies such as high-yielding varieties of cereals, particularly dwarf wheat and rice, and the widespread use of chemical fertilizers, pesticides, and controlled irrigation. As crops began to reach the maximum improvement possible through selective breeding, genetic modification technologies were developed to allow for continued efforts. At the same time, newer methods of cultivation, including mechanization, were adopted, often as a package of practices to replace traditional agricultural technology. This was often in conjunction with loans conditional on policy changes being made by the developing nations adopting them, such as privatizing fertilizer manufacture and distribution. The early dramatic successes of the Green Revolution were in Mexico and the Indian subcontinent (Various, 2024a).

A key leader was agricultural scientist Norman E. Borlaug (1914–2009), the "Father of the Green Revolution". Borlaug studied plant biology and forestry at the University of Minnesota and earned a Ph.D. in plant pathology there in 1942. He began working with the DuPont Company in 1942 but was soon recruited as a research scientist in charge of wheat improvement for the Rockefeller Foundation's Cooperative Mexican Agricultural Program in Mexico, where he worked from 1944 to 1960. Seeking to assist impoverished farmers who struggled with diseased and low-producing crops, Borlaug experimented with novel varieties of wheat, creating disease-resistant strains that could withstand the harsh climate. That work was founded on earlier discoveries of ways to induce genetic mutations in plants, and his methods led to modern plant breeding. In constant demand as a consultant, Borlaug served on numerous committees and advisory panels on agriculture, population control, and renewable resources. He also taught at Texas A&M University (1984–2009), where the Norman Borlaug Institute for International Agriculture was established in 2006. His numerous other honors include the Nobel Peace Prize (1970), the Presidential Medal of Freedom (1977), the National Medal of Science (2004), the Congressional Gold Medal (2006), and the United Nations FAO Agricola Medal (2010) (Pallardy, 2024).

Another important scientific figure was Yuan Longping (1929–2021). Yuan, an agronomist and inventor, was a member of the Chinese Academy of Engineering. He is known for developing the first hybrid rice varieties in the 1970s, part of the Green Revolution in agriculture. Hybrid rice has since been grown in dozens of countries in Africa, America, and Asia—boosting food security and providing a robust food source in areas with a high risk of famine. The technology allowed

China to sustain 20% of the global population on 9% of global arable land. Yuan's work on hybrid rice varieties is credited with saving many lives, and for his contributions, Yuan is known as the "Father of Hybrid Rice". For his achievements, he was awarded the State Preeminent Science and Technology Award of China (2000), the World Food Prize (2004), and the Wolf Prize in Agriculture (2004) (Various, 2024b).

The end result was that the technological developments by Borlaug, Yuan, and many other scientists, combined with favorable governmental policies and increased funding, led to a rapid increase in agricultural output as measured in yields per hectare. Between 1960 and 2000, yields for all developing countries rose 208% for wheat, 109% for rice, 157% for maize, 78% for potatoes, and 36% for cassava. The widespread adoption of Green Revolution technologies led to a significant shift in the food supply function, contributing to a fall in real food prices. Between 1960 and 1990, the food supply in developing countries increased by 12–13% (Pingali, 2012). It is estimated that the Green Revolution saved over a billion people from starvation!

8.1.2 *Opportunity for Pharmaceutical Scientists: Understand the Issues and Own the Problem*

The solutions provided by the Green Revolution helped in solving an existential problem of food for humankind. We, as pharmaceutical scientists, also have an opportunity to solve the existential problem of disease similarly. However, it is very important for all in product development to realize that the choices made during product design may lead to increased costs for the patients, which in turn may prohibit access to medications for a large segment of the global population. It is also acknowledged that the high costs of pharmaceuticals are not always the result of product design choices but are a combination of numerous factors such as corporate strategy, potential patent expirations, market dynamics, cost-benefit analysis, and other health technology assessments. However, every step towards gaining cost efficiency is a worthy step in the right direction to improve global health equity. It is, therefore, helpful to revisit the Innovation Matrix to see how pharmaceutical scientists can help bring some balance to the global health inequity.

Traditionally, product development scientists in the pharmaceutical industry do not have a customer-facing career. There is significant interaction between CMC team members with clinical trial management and the clinical sciences team. However, it is rare for a CMC-based scientist to have a day-to-day interaction with the patient, and therefore, from professional training, there may not be much awareness of the issues faced by the patients in terms of access to medication and costs. In the context of the Innovation Matrix, if the problem of global health inequity is well defined, then the role that pharmaceutical scientists can play in it is not well defined. This puts pharmaceutical scientists' work in the "Breakthrough Innovation"

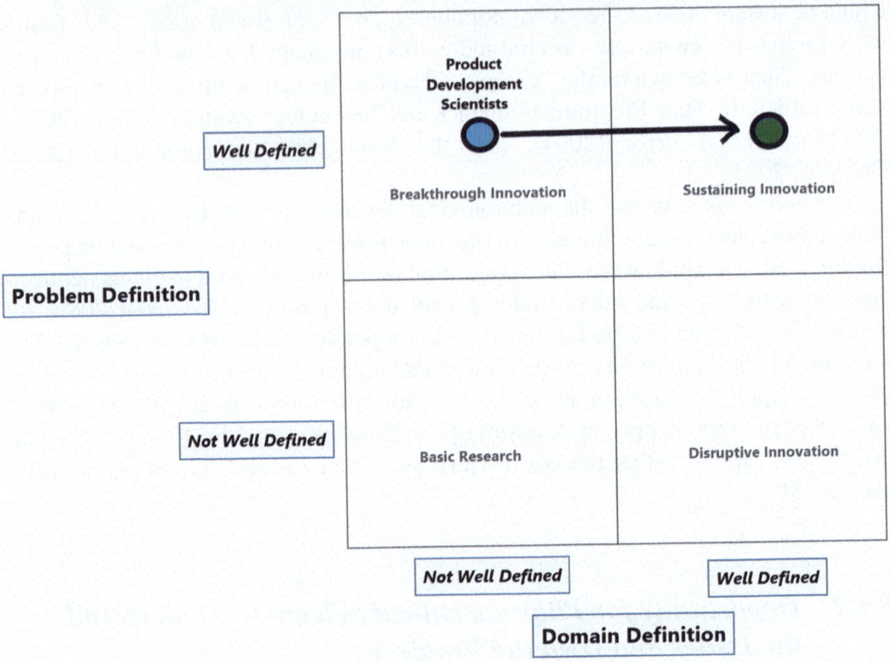

Fig. 8.2 The innovation matrix for alleviating global health inequality that is inclusive of product development scientists

quadrant. This means that unless pharmaceutical scientists develop something that is novel and groundbreaking, their impact on the access to medication and cost aspects of pharmaceuticals may be limited. It is my supposition that by raising the awareness of the product development scientists, we can make them part of the solution and help them to move towards the "Sustaining Innovation" quadrant, as shown in Fig. 8.2.

For example, if the awareness of the costs paid by the patients was factored into product development and was indeed a design parameter that would be scrutinized similar to how other design parameters (such as assay, impurities, shelf life, etc.) are integrated into the target product profile (TPP) and quality target product profile (QTPP), then the overall attempts to bring balance to the global health inequity will be lot more wholesome and collectively owned by the designers of the product. Such an inclusive approach may also help improve patient-focused drug development efforts.

Could it be that in a future state of product development, drug product pricing and reimbursement may be an integral element of the QTPP?

8.2 Legal and Regulatory Efforts to Reduce Costs of Prescription Medicines

The high costs of prescription medicines are a major societal and political issue. These high costs threaten healthcare budgets and limit funding available for other areas in which public investment is needed. This topic also elicits a strongly polarized response from various stakeholders. Proposed solutions discussed in the literature include patent reform, compulsory licensing, transparency in pharmaceutical lobby spending, cap on price increases, and physician-led reforms (Vincent Rajkumar, 2020). Other thought leaders also believe that the United States needs a new strategy for drug innovation that recognizes the new reality of the biomedical innovation process by reinvesting a substantial fraction of the cost savings from Medicare reform into the most powerful engine of innovation in drug development: the NIH and its academic research partners. This would likely generate a flood of truly innovative, life-saving products—not a flock of me-too drugs—that small companies could perfect and Big Pharma could steer to market (Blumenthal et al., 2021). All these opinions (and many others) are important as they allow for a healthy debate that also leads to actionable legislation, policy, and commitments. A few of such ongoing efforts are discussed below.

8.2.1 Changes in Medicare Due to the Inflation Reduction Act of 2022

On 16 August 2022, President Joe Biden signed into law the Inflation Reduction Act of 2022 (IRA). The IRA is a landmark United States federal law that aims to curb inflation by reducing the federal government budget deficit, lowering prescription drug prices, and investing in domestic energy production while promoting clean energy. The IRA enacted the most significant reform for the payment of drugs and biologicals. The IRA's Medicare-related provisions fall into two general categories: (1) reduce prescription drug prices, and (2) reduce beneficiary cost sharing and premiums (Table 8.1) (Goldman et al., 2023). It is expected that over the 10 years between 2022 and 2031, through prescription drug negotiation and the Medicare inflationary rebates, approximately $101 billion and $71 billion, respectively, can be saved. The IRA incorporates many of the drug pricing concepts, and its most crucial healthcare provisions include (Niazi, 2022):

- Establishing a new program for Medicare to directly negotiate prices with pharmaceutical manufacturers for certain high-spend Medicare drugs, with stiff penalties for companies that refuse. Medicare Part D (retail prescription drugs) and Part B (administered by physicians) drug spending is highly concentrated among a relatively small share of covered drugs, mainly those without generic or biosimilar competitors. The number of negotiated drugs is limited to 10 Part D drugs

Table 8.1 Summary of IRA's prescription drug provisions (Goldman et al., 2023)

Aim	IRA provision for medicare	Provision description
Reducing pharmaceutical prices	Introduces drug price negotiations	The federal government must negotiate prices for high-priced, small-molecule, single-source drugs and biologics covered by Medicare and approved by the FDA for more than 9 and 13 years, respectively
	Penalizes price increases	Drug manufacturers will have to pay a rebate if the prices of their single-source drugs (used by Medicare beneficiaries) exceed that of the inflation-adjusted price of the drugs that year. In other words, drug manufacturers must pay a rebate if their drug prices increase more than the inflation rate of the broader economy
	Expands required discounts	Drug manufacturers must pay discounts of 10% during the initial coverage phase and 20% in the catastrophic coverage phase for brand-name medications
Reducing patient cost sharing and premiums	Caps out-of-pocket spending	By eliminating the 5% coinsurance for Medicare Part D catastrophic coverage in 2024 and enforcing an annual $2,000 out-of-pocket spending cap for prescription drug costs covered by Medicare in 2025, Medicare patients will have a hard out-of-pocket maximum similar to many patients with commercial insurance
	Expands eligibility for Low-Income Subsidy (LIS)	The IRA expands the eligibility for full LIS benefits to individuals with incomes between 135% and 150% of the federal poverty level and with resources up to $9,900 for individuals and $15,600 for couples in 2022
	Eliminates vaccine cost sharing	Cost sharing has been eliminated for adult vaccines covered under Medicare Part D
	Limits patient cost sharing for insulin products	Starting in 2023, monthly copayments for insulin products covered under Medicare Part D will be limited to $35. Furthermore, for insulin products administered via a traditional pump and thus covered under Medicare Part B's durable medical equipment benefit, no deductibles can be enforced in addition to the mentioned cap in copayments
	Limits medicare Part D premium increases for beneficiaries	The IRA limits annual increases in Part D base premiums to 6% per year between 2024 and 2029

in 2026, another 15 Part D drugs in 2027, another 15 Part B and Part D drugs in 2028, and another 20 Part B and Part D drugs in 2029 and later years (Cubanski et al., 2023).

- Requiring manufacturers to pay rebates on drugs reimbursed under Medicare Parts B or D for which average (i.e., net) prices increase faster than inflation.

Negotiating drug prices is not new for federal agencies. For example, as with many federal programs, the US Department of Veterans Affairs (VA) has statutorily mandated access to favorable drug pricing. VA provides prescription drug coverage for approximately 9 million enrolled veterans through its outpatient medical clinics and mail-order pharmacy program—the Consolidated Mail Outpatient Pharmacy, or CMOP. VA has significant control over its prescription drug program—a singular purchaser with a unified national formulary. In general, medications must be prescribed by a VA provider, filled at a VA pharmacy, and listed on the national formulary. VA also fills prescriptions from authorized non-VA providers (i.e., in the Community Care setting). VA may provide nonformulary drugs in cases of medical necessity. VA primarily uses a direct purchase approach to acquire drugs. VA uses the pharmaceutical prime vendor program—a single vendor awarded by contract—to acquire drugs directly from manufacturers for distribution through its facilities. After distributing the drugs, the prime vendor does a "charge-back" to the manufacturer to obtain the difference between its purchase price and the VA-negotiated sales price. As such, VA's prices are generally not subject to markups, and its prices do not include costs for storage, overhead, or dispensing (GAO, 2020).

In a report released by the US Government Accountability Office (GAO), it was assessed that the VA paid, on average, 54% less per unit for a sample of 399[1] brand-name and generic prescription drugs in 2017, as did Medicare Part D even after accounting for applicable rebates and price concessions in the Part D program. GAO also found that 233 of the 399 drugs in the sample were at least 50% cheaper in VA than in Medicare, and 106 drugs were at least 75% cheaper. Only 43 drugs were cheaper in Medicare than in VA. The percent difference in price between the two programs was greater on average for generic drugs. Specifically, VA's prices were 68% lower than Medicare prices for the 203 generic drugs (an average difference of $0.19 per unit) and 49 percent lower for the 196 brand-name drugs (an average difference of $4.11 per unit) (GAO, 2020). The GAO report ascertains that one of the reasons (apart from the VA's access to statutory discount on its drug purchases—not available to Part D plans) for this discrepancy is that while Medicare Part D is a much larger program overall than VA, the unified structure of VA may provide additional leverage in negotiations with drug manufacturers. For example, Medicare Part D's 42.5 million beneficiaries are divided among numerous prescription drug plans. In December 2017, the largest plan had approximately 5.6 million enrollees, with most having an enrollment of under 1 million individuals. In contrast to Medicare Part D, the VA represents a single unified plan capable of acting for all its enrolled veterans; thus, in effect, it had an enrollment of 9 million individuals in 2017. Consequently, VA's large purchasing volume may strengthen its bargaining position when negotiating (GAO, 2020) (Fig. 8.3).

Therefore, for the first time, the law provides Medicare the ability to directly negotiate the prices of certain high-expenditure, single-source drugs without generic or biosimilar competition. On August 29, 2023, CMS announced the ten drugs covered under Medicare Part D that were selected for the first cycle of negotiations

[1] GAO's sample of 399 drugs included the top 100 brand-name and generic drugs in Medicare Part D in 2017, by: (1) highest expenditures; (2) highest utilization (by quantities dispensed); and (3) highest cost-per use.

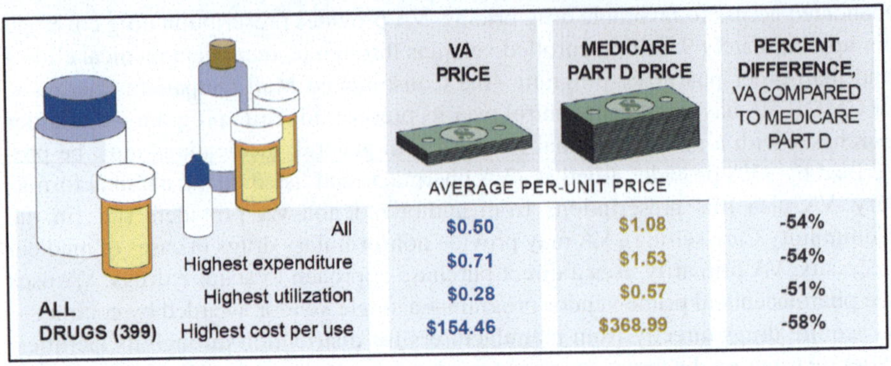

Fig. 8.3 Comparison of prices paid by VA vs. medicare part D for selected drugs in 2017. (GAO, 2020)

(Table 8.2). The selection of drugs for the first round of negotiations was made through a series of inclusion criteria, such as identifying potential qualifying single-source drugs[2] and drugs with the highest total Part D gross covered prescription drug costs. The companies that manufacture these ten drugs have agreed to participate in the CMS Negotiation Program. The IRA relating to drug price reduction applies only to Medicare, which serves 20% of patients. The IRA does not address cost savings for the other 80% of patients in the private market. However, it can be anticipated that future changes in the law could address further cost savings.

8.2.2 *Support Competitive Marketplace for Generics and Biosimilars*

Recently, the FDA has established policies that support a competitive marketplace for generic drugs and biosimilar products, which can lower prices. In 2017, the FDA announced the Drug Competition Action Plan (DCAP) further to encourage robust and timely market competition for generic drugs and help bring greater efficiency and transparency to the generic drug review process without sacrificing the scientific rigor underlying our generic drug program. Through this plan, the FDA is helping remove barriers to generic drug development and market entry in an effort to spur competition so that consumers can get access to the medicines they need (FDA, 2024). Similarly, in July 2018, the FDA introduced the Biosimilar Action Plan (BAP) to help stimulate the US biosimilar market (FDA, 2018). Through the DCAP

[2] Single-source drugs are drugs for which at least 7 years, or biologics for which at least 11 years, have elapsed between the FDA approval or licensure and the selected drug publication date, and for which there is no generic or biosimilar competition.

Table 8.2 Drugs selected for first cycle of medicare drug price negotiation program (CMS, 2023)

Drug name	Participating manufacturer
Eliquis	Bristol Myers Squibb
Jardiance	Boehringer Ingelheim
Xarelto	Janssen Pharmaceuticals, Inc.
Januvia	Merck
Farxiga	AstraZeneca
Entresto	Novartis
Enbrel	Amgen
Imbruvica	Pharmacyclics
Stelara	Jannsen Biotech, Inc.
Fiasp; Fiasp FlexTouch; Fiasp PenFill; NovoLog; NovoLog FlexPen; NovoLog PenFill	Novo Nordisk

and BAP approach, the FDA is working to help remove barriers to generic and biosimilar market entry, including through workshops and guidance to facilitate product development, initiatives to improve the quality of applications and enhance the efficiency of review by the FDA; and efforts to reduce monopolistic strategies that delay competition in the marketplace. The agency has issued policy documents designed to ensure timelier generic drug approvals and earlier patient access by reducing the number of assessment cycles, facilitating prompt labeling updates, and providing advice on how to avoid common application deficiencies that lead to approval delays (Bush & Sommers, 2022).

8.3 Sustainable Value Creation Opportunities

The pharmaceutical industry has an image problem. Public distrust in the pharmaceutical industry has increased, in part due to perceptions of pharmaceutical manufacturers as profit-seeking and in part due to their actions, including off-label marketing, overcharging government programs, and concealing data. Notwithstanding the underlying mechanism, an unfavorable public perception of pharmaceutical manufacturers is concerning if it translates into poor medication adherence, lack of participation in clinical trials, and rejection of effective health interventions, including vaccine campaigns (Singh et al., 2023). A Gallup Poll conducted in August 2019 found that 58% of Americans held negative views of the pharmaceutical industry, while only 27% held positive views of it. It is the lowest the industry has ever been ranked in the poll, which began in 2001. Gallup noted that high drug prices, the opioid epidemic, and Big Pharma's big lobbying efforts are all factors that likely played into respondents' frustration with the industry (Harvard, 2019). As a member of the pharmaceutical industry, I find such an assessment of our industry unfair. With the difficulty and challenges associated with drug discovery and development, it is commendable that numerous highly qualified scientists work very hard to develop medicines that can help improve the quality of life of numerous patients. The pharmaceutical industry is not a monolithic entity but a symbiotic organization where numerous disciplines have to come together to develop something novel. Such innovative work is not easy and there is no guarantee of success either. But, alas, perception is reality!

8.3.1 Integrated Healthcare Networks to Improve Value Creation

The lack of success in reversing this perception may be due to the impression that the steps taken so far are mostly 'cosmetic'. They are performed by departments such as 'public relations' or 'corporate social responsibility' and are, by nature, decoupled from a company's commercial operations. The key issue is that the revenues generated by pharmaceutical companies are not perceived to match the contribution they make to the individual patient and to the healthcare system as a whole (Sobrio & Keller, 2007). This disconnect with the stakeholders is often underemphasized as the industry defends itself by saying that pharmaceutical R&D is expensive and, therefore, the costs of products coming out from said R&D will be high, too. Nevertheless, for a general member of the public, this argument is becoming stale, cliched, and hackneyed. As different stakeholder groups become more interrelated, influence and decision patterns in the healthcare environment change.

As discussed in Sect. 3.6, there is growing interest in research that involves patients and other stakeholders as partners in study planning, conducting, and sharing information about health research results. Patient-reported outcome measures (PROMs) are one strategy to ensure that patient's perspectives are systematically incorporated into the approaches to delivering healthcare services and valuing the performance of the healthcare system. Another approach could be to develop integrated healthcare networks that improve value creation. In such an approach, the importance of a steady revenue stream and return on investment (ROI) for shareholders is not discounted but is integrated into the operational model. Such a balanced approach to pharmaceutical value creation in networked healthcare is visualized in Fig. 8.4.

There are three main dimensions for which a company creates sustainable value: (1) financial performance (i.e., in terms of its own profitability), (2) healthcare effectiveness (i.e., in terms of concrete patient treatment outcomes), and (3) healthcare efficiency (i.e., in terms of contribution to cost-containment in the healthcare system). Taking the network perspective, the main output of the healthcare system must ultimately be 'health', measured and defined in patient outcomes. This perspective, however, needs to be balanced with the cost at which this output is delivered and the profitability levels of all stakeholders operating in the network.

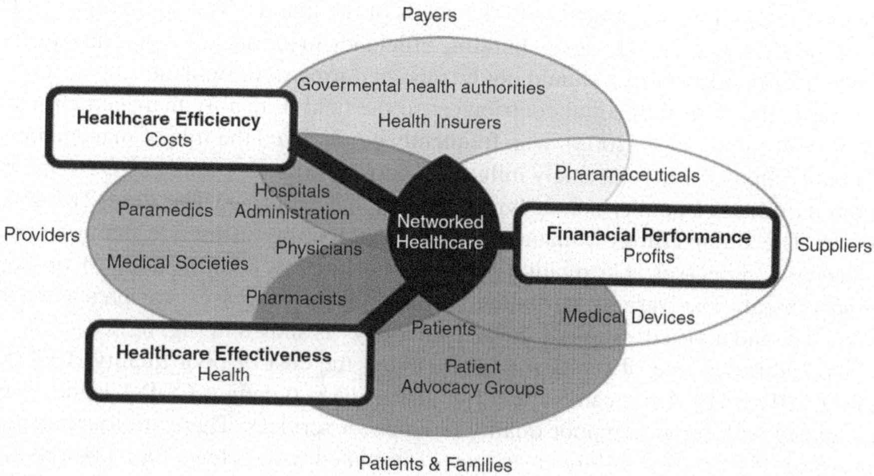

Fig. 8.4 Three areas of balanced pharmaceutical value creation in networked healthcare. (Sobrio & Keller, 2007)

Effectiveness in terms of health provided, efficiency in terms of cost of delivering health, and financial performance in terms of profitability of healthcare operations are of concern to all stakeholders of the healthcare network and should be made transparent to all (Sobrio & Keller, 2007).

8.3.2 *The Juran Trilogy®*

The pharmaceutical industry is very conscious of any development efforts that can help reduce costs and minimize the time it takes for the products to reach the commercial market. Current economic and political pressures are leading many research-based pharmaceutical and biopharmaceutical companies to re-assess decades-old R&D strategies and focus their efforts on improving efficiency, boosting output, and meeting unmet patient needs. As discussed in Sect. 5.3, the focus on targeted medicines, often addressing the needs of relatively small patient populations, is enabled by advances in genomics, proteomics, and other tools that allow researchers to understand disease mechanisms better and identify patient subpopulations. There are potential cost-saving strategies for clinical trials, too, that allow a company to allocate and prioritize its resources better. Enhanced drug discovery and efficient clinical trials alone, however, are not enough to help the industry remain competitive in today's challenging pharmaceutical marketplace. Companies must also address the cumbersome, expensive, and inefficient drug development process itself. Inefficiencies in drug development can stem from unoptimized drug substance and drug product manufacturing processes leading to low yields and high costs of goods, lengthy stability studies, underestimating the needs of patient compliance, and misalignment with the needs of the market. The adoption of QbD is a good starting point. However, building efficiency in product design is an equally important opportunity that should not be missed during development.

One of the most influential contributors to the field of quality management was Dr. Joseph Juran (1904–2008), who frequently emphasized the role of management in quality. In one of his extremely influential books, *Juran on Quality by Design*, Dr. Juran introduces a quality management concept that he termed The Juran Trilogy®. As per The Juran Trilogy®, managing quality is done by using a system of three managerial processes: (1) quality planning, (2) quality control, and (3) quality improvement. The various attributes for these three processes are presented in Table 8.3, and a visualization of the Juran Trilogy® is shown in Fig. 8.5.

Dr. Juran was one of the first to write about the cost of poor quality (COPQ) (ASQ, 2018). The American Society for Quality (ASQ) defines COPQ as the costs associated with providing poor-quality products or services. There are four categories of COPQ: internal failure costs (costs associated with defects found before the customer receives the product or service), external failure costs (costs associated with defects found after the customer receives the product or service), appraisal costs (costs incurred to determine the degree of conformance to quality requirements), and prevention costs (costs incurred to keep failure and appraisal costs to a

Table 8.3 The Juran Trilogy (Juran, 1992) (Reprinted with permission from Elsevier)

Process	Attributes
Quality planning	This is the activity of developing the products and processes required to meet customer's needs. It involves a series of universal steps, which can be abbreviated as follows: Establish quality goals Identify the customers—those who will be impacted by the efforts to meet the goals Determine the customer's needs Develop product features that respond to customer's needs Develop processes that can produce those product features Establish process controls and transfer the resulting plans to the operating forces
Quality control	This process consists of the following steps: Evaluate actual quality performance Compare actual performance to quality goals Act on the difference
Quality improvement	This process is the means of raising quality performance to unprecedented levels ("breakthrough"). The methodology consists of a series of universal steps: Establish the infrastructure needed to secure annual quality improvement Identify the specific needs for improvement—the improvement projects For each project, establish a project team with clear responsibility for bringing the project to a successful conclusion Provide the resources, motivation, and training needed by the teams to Diagnose the causes Stimulate the establishment of remedies Establish controls to hold the gains

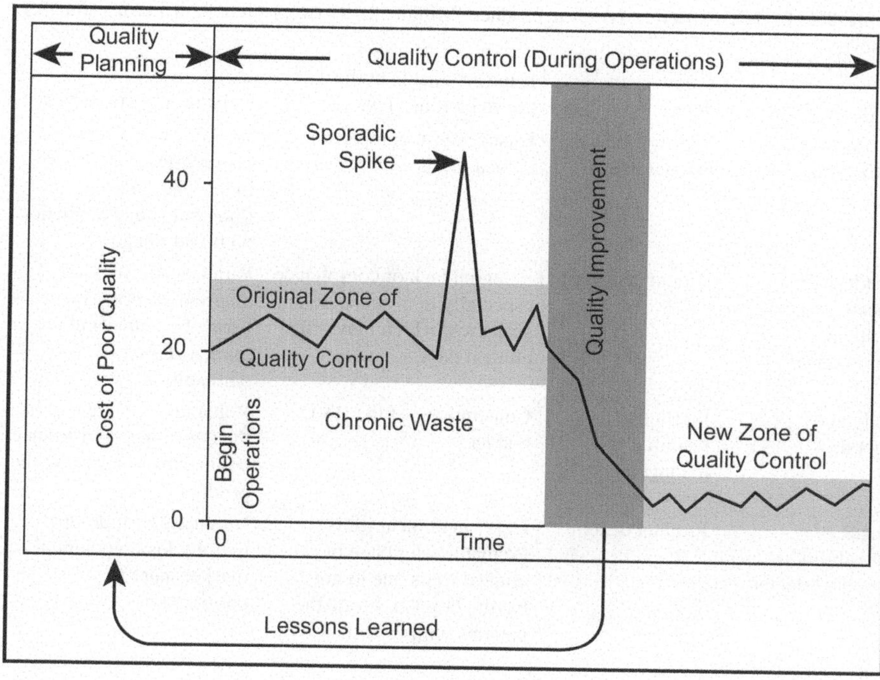

Fig. 8.5 The Juran Trilogy® diagram. (Juran, 1992) (reprinted with permission from Elsevier)

minimum). Cost of quality is a methodology that allows an organization to determine the extent to which its resources are used for activities that prevent poor quality, which appraise the quality of the organization's products or services and that result from internal and external failures. Having such information allows an organization to determine the potential savings to be gained by implementing process improvements.

8.3.3 *Quality by Efficient Design (QbED)*

Dr. Juran proclaims that many quality crises and problems have been traced to the way in which quality was planned in the first place. Therefore, in a way, our quality problems have been planned that way (Juran, 1992). This is a profound statement as one takes the time to examine it. On the surface, it implies that our quality problems are not extraneous events but are an integral part of the design process itself. In a more nuanced interpretation, it can also imply that the failure on the part of a product's designer to account for potential issues during the design phase fundamentally locks in product defects. These defects have to be eventually managed as part of the

Table 8.4 Impact of defect type on product (Mittal, 2019) (Reprinted with permission from Elsevier)

Type of defect	Category of defect	If the product is in the pre-marketing stage	If the product is marketed
Low inventory or stock-out	Supply chain management	Unsuccessful completion of clinical trials	Loss of market share to competition Loss of revenue Potential negative impact on brand image
Failure to account for patient preferences	Patient centricity	Potential lack of compliance, especially in at-home dosing scenarios, which may affect clinical data	Patients may look for alternative choices Loss of revenue and market share to competition
High cost of goods	Inefficient resource management	Constraining of the R&D budget	Reluctance on the part of the consumer or insurance companies to purchase the product
Lack of flexibility in product dosing	Product design	Delay in clinical trials Premature stoppage of clinical trials due to not having doses to titrate the patients with	Limited use of product, thereby losing potential market share to competition

product's life cycle, which can be quite costly. In other words, the failure of imagination on the part of the product designer is the potential root cause of the defects! Therefore, early anticipation and identification of potential defects (Table 8.4) can help increase the product's quality and reduce chronic waste (Mittal, 2019).

On a similar note, a discussion on chronic waste is equally interesting. Waste, by definition, is any substance that is discarded after its primary use or is worthless, defective, and of no use. We all implicitly understand the need to reduce waste and use all of our resources judiciously. Unfortunately, chronic waste is a problem in all manufacturing industries, and pharmaceuticals are no different. However, just because chronic waste is inherent in manufacturing, it does not mean that steps cannot be taken to reduce it. There are two general categories of sources that lead to chronic waste, as discussed in Table 8.5. Efficiency in product design is an important principle whose impact on a company's future marketplace competitiveness may not be as evident at an early stage but becomes increasingly prominent as the drug advances in clinical trials. Therefore, an integration of QbD principles with typical late-stage concepts of patient centricity, manufacturing excellence, business acuity, etc., during early design stages can improve the efficiency of the manufacturing processes and economic viability of the product. This is the ultimate intent of Quality by Efficient Design (QbED) (Fig. 8.6).

Table 8.5 Examples of chronic waste in pharmaceuticals (Mittal, 2019) (Reprinted with permission from Elsevier)

Source type	Contribution to chronic waste	Examples
Waste by design	Permanent unless steps are taken to "design it out" from the product	• **Product design:** • Designing a formulation that does not take into account patient centricity and brand enhancement • Failure to account for future scale-up issues in the product's design • Incorporating marketing input too late into product design • **Manufacturing process:** • Choosing too small or too large of a batch size that requires numerous resupply campaigns or waste material that is not needed • Choosing a manufacturing process that inherently produces a low yield of product • There is not enough shelf life for the product due to either the molecule's instability or poor formulation/manufacturing choices • Choosing a dosage form that is not amenable to scale-up principles • Failure to incorporate multivariate analysis and statistical process control concepts to evaluate the product's critical quality attribute
Waste by poor utilization	Temporary and could be addressed by better communication with clinical trials design team	• **Overambitious timelines:** • Failure to account for primary packaging, kitting, and labeling into timelines that shorten the product's useable shelf life • **Sub-optimal clinical trial designs:** • Poor ratio of patients/sites ratio would complicate supply chain coordination and would burden existing supplies, leading to unnecessary remanufacturing • Designing a large number of double-blind studies that require over-encapsulation of product, leading to extended manufacturing and testing timelines before clinical dosing

8.4 Summary—Where Do We Go from Here?

The hope is that this book is a critical reminder of why it is important for pharmaceutical scientists to keep an eye on integrating (global) market considerations into drug discovery, product design, and process development. Drug discovery is the exciting initiation of a development program that we hope will ultimately deliver a new medicine of real therapeutic value to the patient in need. In the 21st century, the science of drug development is an established field that requires a dedicated understanding of several disciplines and entails a symbiotic partnership between numerous subject matter experts. Given the experience that we now have in R&D, the

Fig. 8.6 Quality by efficient design (QbED). (Mittal, 2019) (Reprinted with permission from Elsevier)

steps taken toward establishing a drug's safety and efficacy, product development, and commercialization have long been standardized. However, the development of pharmaceutical products is not immune to the irrefutable laws of economics.

As we reviewed some of the downstream financial, competitive, and prescribing pressures that ultimately determine whether a new drug will ever reach the market, the challenge to maintain the innovation potential continues to be formidable. For the very few drugs that make it to the market, it is extremely critical that their value proposition is not diminished because of the high costs of medicines for the patients. The goal of this book is to raise awareness around strategic decision-making during drug development that allows for a pharmaceutical asset to provide economic benefits to innovative companies without compromising quality and therapeutic benefits to patients. As everyone knows, the development of a robust drug product requires juggling many competing priorities, such as overcoming scientific challenges, following regulatory requirements, and managing business-related concerns. However, it is equally important to be aware of the impact of development time and cost of goods on patent protection, exclusivity, marketability, and competitiveness.

It is, therefore, critical for pharmaceutical scientists to realize the opportunities that can help in not only building back trust in our industry but also taking proactive steps that enhance the value proposition of our products. After all, we are genuinely passionate about bringing new therapies to the marketplace and providing better health outcomes to our patients. ***Aren't we?***

References

ASQ. (2018). *Joseph M. Juran: A search for universal principles*. Retrieved September 13 from https://asq.org/about-asq/honorary-members/juran

Blumenthal, D., Miller, M. E., & Gustafsson, L. (2021, October 1, 2021). The U.S. can lower drug prices without sacrificing innovation. *Harvard Business Review*. https://hbr.org/2021/10/the-u-s-can-lower-drug-prices-without-sacrificing-innovation

Bush, L., & Sommers, B. D. (2022). Improving prescription drug affordability through regulatory action. *JAMA Health Forum, 3*(8), e223180. https://doi.org/10.1001/jamahealthforum.2022.3180

CMS. (2023). *Medicare drug price negotiation program*. Retrieved from https://www.cms.gov/files/document/fact-sheet-manufacturer-agreements-selected-drugs-ipay-2026.pdf

Cubanski, J., Neuman, T., Freed, M., & Damico, A. (2023). *How will the prescription drug provisions in the inflation reduction act affect Medicare beneficiaries?* KFF. Retrieved October 23, 2023 from https://www.kff.org/medicare/issue-brief/how-will-the-prescription-drug-provisions-in-the-inflation-reduction-act-affect-medicare-beneficiaries/#

FDA. (2018). *Biosimilars action plan*. FDA Retrieved from https://www.fda.gov/media/114574/download

FDA. (2024). *FDA drug competition action plan*. FDA. Retrieved March 1, 2024 from https://www.fda.gov/drugs/guidance-compliance-regulatory-information/fda-drug-competition-action-plan

GAO. (2020). *Report # GAO-21-111: "Prescription drugs: Department of veterans affairs paid about half as much as medicare part D for selected drugs in 2017"*. Retrieved from https://www.gao.gov/assets/gao-21-111.pdf

Goldman, D., Grogan, J., Lakdawalla, D., Liden, B., Shafrin, J., Than, K.-S., & Trish, E. (2023). *Mitigating the inflation reduction act's adverse impacts on the prescription drug market*. https://healthpolicy.usc.edu/research/mitigating-the-inflation-reduction-acts-potential-adverse-impacts-on-the-prescription-drug-market/

Harvard. (2019). *Poll shows Americans are fed up with pharmaceutical industry*. Harvard T.H. Chan School of Public Health. Retrieved March 1, 2024 from https://www.hsph.harvard.edu/news/hsph-in-the-news/poll-shows-americans-are-fed-up-with-pharmaceutical-industry/

Juran, J. M. (1992). *Juran on quality by design: The new steps for planning quality into goods and services*. Free Press.

Mittal, B. (2019). How to integrate quality by efficient design (QbED). In *Product development*. Elsevier Academic Press.

Niazi, S. K. (2022). The inflation reduction act: A boon for the generic and biosimilar industry. *Journal of Clinical Pharmacy and Therapeutics, 47*(11), 1738–1751. https://doi.org/10.1111/jcpt.13783

Pallardy, R. (2024). Norman Ernest Borlaug. In *Encyclopedia Britannica*.

Pingali, P. L. (2012). Green revolution: Impacts, limits, and the path ahead. *Proceedings of the National Academy of Sciences, 109*(31), 12302–12308. https://doi.org/10.1073/pnas.0912953109

Satell, G. (2017). *Mapping innovation: A playbook for navigating a disruptive age*. McGraw Hill Education.

Singh, Y., Eisenberg, M. D., & Sood, N. (2023). Factors associated with public trust in pharmaceutical manufacturers. *JAMA Network Open, 6*(3), e233002. https://doi.org/10.1001/jamanetworkopen.2023.3002

Sobrio, A., & Keller, M. (2007). Integration through sustainable value creation. *Journal of Medical Marketing, 7*(2), 102–113.

Various. (2024a). *Green revolution*. Wikipedia. Retrieved February 28, 2024 from https://en.wikipedia.org/wiki/Green_Revolution

Various. (2024b). *Yuan Longping*. Wikipedia. Retrieved February 28, 2024 from https://en.wikipedia.org/wiki/Yuan_Longping

Vincent Rajkumar, S. (2020). The high cost of prescription drugs: Causes and solutions. *Blood Cancer Journal, 10*(6), 71. https://doi.org/10.1038/s41408-020-0338-x

Index

B. Mittal, *Sustainable Global Health Systems and Pharmaceutical Development*, AAPS Advances in the Pharmaceutical Sciences Series 60,
https://doi.org/10.1007/978-3-031-50415-0

The manufacturer's authorised representative in the EU is Springer Nature Customer Service Centre GmbH, Europaplatz 3, 69115 Heidelberg, Germany. If you have any concerns regarding our products, please contact ProductSafety@springernature.com

Printed and bound by CPI Group (UK) Ltd, Croydon, CR0 4YY
23/09/2026
02221546-0005